Humanity's Mirror

Humanity's Mirror

150 Years of Anatomy in Melbourne

Ross L Jones

HADDINGTON PRESS

First published in 2007 by Department of Anatomy and Cell Biology,
The University of Melbourne.

Produced by Haddington Press Pty Ltd
Box 3182, Domain Road P.O.
South Yarra, Victoria 3141

Phone: (61 3) 9866 2758
Fax: (61 3) 9866 2751
Email: haddingtonpress@bigpond.com
Website: www.haddingtonpress.com.au

The National Library of Australia Cataloguing-in-Publication entry:

Jones, Ross L., 1955- .
Humanity's mirror : 150 years of anatomy in Melbourne.

Bibliography.
Includes index.
ISBN 9780646473000 (hbk.).

1. Anatomy - Research - Victoria - Melbourne - History. I.
University of Melbourne. Dept. of Anatomy and Cell Biology.
II. Title.

571.3072099451

Text and cover design by Andrew Cunningham – Studio Pazzo
Printed by Everbest in China

10 9 8 7 6 5 4 3 2 1

DEDICATION

For Susan and Wilma

Foreword

How old is the Department of Anatomy and Cell Biology at the University of Melbourne? In the narrowest sense, the current Department is young, as its name was changed from Anatomy to Anatomy and Cell Biology as recently as 1993. Its parent, the Anatomy Department, was formed in 1906 when the Department of Descriptive and Surgical Anatomy and Pathology split, and Richard Berry was appointed as the first professor of the free-standing Anatomy Department. Thus, 2006 may be regarded as the centenary of our Department. However, our history extends further back in time. George Halford was appointed Professor of Anatomy, Physiology and Pathology in 1862—the first professor in the Medical School. Even prior to this, Anatomy had been taught in the University of Melbourne since its incorporation in 1853. Thus, Anatomy has been flourishing here for over one hundred and fifty years.

To commemorate our centenary we decided to publish a history of the Department. A committee was formed consisting of current members of the Department (Christopher Briggs, John Furness and Antony Goodwin) and previous members (Ian Darian-Smith, James Guest, Geoffrey Kenny and Graeme Ryan). The committee was fully cognisant of the rich history of the Department. The early professors' interests and influence extended well beyond the Medical School and beyond the University; they played a major role in Melbourne's political and social milieu. We were also aware that parts of our history were, at times, controversial. Members of the Department wanted an interesting book that detailed our history in a balanced and objective manner. Therefore, it was decided that a professional historian with expertise in the subject matter was called for. Melbourne medical historian, Ross L Jones, has a long-standing interest in the history of a number of key figures in the Department's early life. We are fortunate that Ross agreed to write our history, and indeed welcomed the opportunity. He has committed a great deal of his own time to researching the material and we are indebted to his energy, enthusiasm and integrity.

As with any such work, it is relatively easy to paint an accurate picture of the Department until about thirty years ago. The more recent activities of the Department, which are changing rapidly, are more difficult to encapsulate. In addition, in its early days the Department was small and staff numbers were stable. Today the Department has over ninety employees, many of them funded from research grants, and turnover is significant. It is not feasible to give an account of all those who have passed through the Department in recent times. Therefore, a decision was made that the book would concentrate on the period up to the end of Sunderland's era (*circa* 1970s). The last chapter is not meant to be comprehensive; rather it indicates the changes in direction of the Department in recent times.

Antony Goodwin

Head of Department, 2007

Contents

List of illustrations

Acknowledgments

THIS BOOK WOULD not have been completed without the help of many. I wish to pay particular tribute to the committee who guided and offered much support. Ian Darian-Smith, Ian Dillon, John Furness, James Guest, Geoff Kenny and Graeme Ryan gave excellent and detailed feedback, and were always willing to provide expert advice. It was a comforting thought to know that Geoff was always on the other end of the phone. Tricia Hartshorn worked tirelessly on many of the important organisational activities and provided endless cheer. Apart from his support on the committee, Chris Briggs provided invaluable help, especially with the last chapter. The pivotal figure was Antony Goodwin. He got the project rolling, providing excellent and positive leadership along the way and saw it through to its conclusion. The book simply would not have been written without his vision. Others have provided feedback on various aspects of the draft or provided critical support along the way. This group includes Daine Alcorn, Warwick Anderson, Ann Brothers, John Collins, Barry Christophers, Nick Hamilton, Keith Henderson, Cecily Hunter, Coralie Kenny, Janet McCalman, Neil Robertson, Dick Selleck, Gordon Keys Smith, Bernice Stratford, Ian Sunderland, John Waller, Anne Westmore and Mary Wheeler. A significant proportion of my work was conducted in archives. Leading me through that labyrinthine world was Joyce Chia and the excellent staff at the University of Melbourne's archives. Collecting and preparing the illustrations would have proved impossible without Stuart Thyer, assisted by Michelle Gough. All at Haddington Press worked tirelessly on the project. In particular, I would like to thank Paul and Georgina McSweeney, Diane Brown, Andrew Cunningham, Meredith Butler, Margaret Findlay and Michael Cunningham. The patience of my sons, Kam, Pat and Ant was invaluable. So too was the support and encouragement from my dear wife, Susan.

The project would not have been possible without the generous support of the Jack Brockhoff Foundation, Mr Tony Buzzard, Professor Stephen Cordner, Dr Marie Cockbill, Mr William Swaney, Professor Brian Tress, Dr Mary Wheeler and, of course, the Department of Anatomy and Cell Biology, The University of Melbourne.

Introduction

WHY *HUMANITY'S MIRROR*? In the comedy *The Brothers*, written in Rome about 160 *BC* by the Roman playwright, Terence, the character Demae says:

> *Inspicere, tanquam in speculum, in vitas omnium iubeo, atque ex aliis sumere exemplum sibi*, or "... look into the lives of all men as into a mirror and thence to take example."[1]

This was chosen as the motto (and hence the source) for the title chosen for the first Medical Students' Society magazine, *Speculum*, in 1884, after a long search in the Baillieu Library at the University of Melbourne. Why choose a name and motto from antiquity? This was a provocative act at the time, as Melbourne medical students felt aggrieved by the image they perceived they had within the University. The first editorial of *Speculum* claimed that the medical student body was known around the University as the 'alleged author of the disturbances at the annual University Commencement.'[2] And those on the 'other side' (all other faculties situated across the University lake) had frequently claimed that a Bachelor of Arts should be made a prerequisite for the medical degree, so as to civilise the medicos with the study of the classics, hence the choice of a title from antiquity. This, of course, creates an amusing ambiguity because a speculum is also an exploratory medical instrument.

Over the years *Speculum* became a thorn in the side of authority and a powerful tool in uniting the medical students. The student magazine launched attacks on all and sundry, including the University Council, major hospital boards and, to their chagrin, occasionally the medical professors themselves. The committee of the Medical Students' Society, in one instance, was granted a private interview with the Premier of Victoria due to its lobbying. Some issues of *Speculum*, in fact, were completely suppressed and heavily censored, although interestingly this was not the case when the Society aggressively championed the Medical School. *Speculum* has been a wonderful source and I would like to acknowledge those anonymous student contributors—now mostly departed.

The feeling of being set apart from the rest of the University arose in the early days of the Medical School, accentuated by its isolated position across the lake. This strong sense of identity remained throughout its history, although this was fuelled by many disparate causes over time. Why had the Medical School been separated from the University? The odour from the dissecting room, it was argued, would upset staff and students who were not inured to it. Thus the study of Anatomy was the reason for the separation. The dead human body as its subject has also meant that there have been many controversial characters and episodes in the Anatomy Department's history. Famous Melbourne criminals, for example, walk in and out of the pages of this book. And although they were only colonials, Melbourne Anatomy professors bowed to no one, not even a dominant figure in science, such as Thomas Henry Huxley. There are moments of near disaster and great triumphs in the history of the Department.

Anatomy has been central to the studies of most students who have passed through the Medical School at the University of Melbourne. The curriculum, however, has changed significantly over nearly one hundred and fifty years. So too has the Anatomy Department, perhaps more than any other. This history tells part of the fascinating story of the evolution that has allowed the Department to survive and prosper, unlike many other anatomy departments in other medical schools. And it is the professors who dominate for related reasons. The archives overwhelmingly record their life and work and few others are as well-represented. At first I was dismayed until I eventually realised why this was so. The Department relied heavily on ex-students who returned to fill positions as lecturers and demonstrators for short periods. So there were few full-time, long serving members of staff other than the

Professor Frederick McCoy (front row, sixth from left), the University's first teacher of anatomy, and a group of medical students, *circa* 1870. Courtesy: Medical History Museum, The University of Melbourne.

professors. Professors Halford, Allen, Berry and Wood Jones were plagued by this problem and Wood Jones resigned. The situation began to change under Sunderland, but the comparative luxury of current staffing levels would have altered the hue of the first four professors—probably to green in the case of Wood Jones and Allen, and red for Halford and Berry. Therefore, what is missing in this work is a full record of the honorary work contributed by many graduates, often unrecorded, and yet essential to the survival of the Department. There is also little mention in the archives of the non-medical anatomy students in their own right, and yet when I refer to the student body, I am referring to physiotherapists (known as masseuses or masseurs in the early days), dentists and science students. Among other omissions are the teaching of Anatomy in places other than the University of Melbourne and those many institutions and individuals that have connections with the Department, most particularly surgeons. There was no room in this history for such fascinating excursions. I was commissioned to write an intellectual and social history of the Department, and accordingly, not every person who has been involved is mentioned. A biographical index, however, is currently being prepared.

When the Department went through a particularly intense evolution after the introduction of the new curriculum at the turn of the last millennium, much that was recognisable in the old Department disappeared. And while changes occurred in the revision of the curriculum in the 1970s, Anatomy emerged in a relatively strong position. The consequences of recent and more radical changes are currently appearing in the form of a debate about whether medical students are given adequate time to study the subject. And this question remains unresolved. The last chapter, therefore, summarises the issues and changes over the last thirty years without any critical analysis. This chapter belongs in the introduction to the sequel that will follow this history of the Department.

Why did the founders of *Speculum* adopt the image of the mirror? Perhaps this was because when studying bodies other than themselves, medical students were encouraged to reflect on their own human qualities, and to engage in inward searching and self-criticism that would inspire empathy and sympathy for their future patients and provide an understanding of all mankind. This was the ideal. Over the almost one hundred and fifty years of dissection in the Department many responses have surfaced from those who stood over the cadavers in the dissecting room. Many, like Richard Berry, the first professor whose designation was simply 'Professor of Anatomy', saw no divine plan and became materialists. Others, like his successor, Frederic Wood Jones, have been inspired to believe in some non-materialist, even spiritual vision of the human world. Thus the study of Anatomy, perhaps more than any other biological science, amounts to more than the sum of its parts.

CHAPTER ONE

Melbourne's anatomists in the nineteenth century: 'the best of friends, and the closest of cronies'

On Saturday 11 May 1901, the University of Melbourne held its Commencement Ceremony in the grand neo-Gothic pile that was Wilson Hall. This annual occasion proved to be somewhat extraordinary, however, as the Council of the University had invited the Duke and Duchess of Cornwall and York as official guests of honour. The Duke (later King George V) had come to Melbourne to open the sitting of the first Federal Parliament in the Royal Exhibition Building in Carlton two days earlier.[1] The Duke of York was not the only dignitary visiting the University for the ceremony. As recognition of the beginning of Australian federal government, the University was to award an honorary Master of Arts to Australia's foundation Prime Minister, Edmund Barton (1849–1920).

The occasion became something of a noisy celebration of the success of Medical School graduates. Among the various songs sung to the accompaniment of stamping feet, it was the medical students' contribution that created the greatest effect. This was on account of the tremendous growth of the School, and on this notable occasion, many medical students (a significant proportion of the total numbers graduating) were being awarded their degrees (MB, MS and MD). By 1897, four years earlier, medical students had, for the first time, comprised fifty per cent of the university population, numbering two hundred and sixty—a radical change from the early days of the School.[2] As the various medical degrees were being awarded at this, the inaugural Commencement Ceremony of the twentieth century, and to the tune of the popular ballad and hymn tune 'Rousseau's dream', the successful medical candidates were encouraged to:

> Saw my ...y leg off, Saw my ...y leg off, Saw my ...y leg off
> SHARP!!!!

According to those present the 'performance came spontaneous and like a thunderclap. The Duchess jumped. The Council looked thunderous. The Duke laughed'[3] and thus confirmed his reputation, at least among the medical graduates, as a man of the people.

By 1901 the School had come a long way from its humble beginnings with three students, no buildings to speak of, and significant hostility to its founding in a city at times oversupplied with medical practitioners.[4] It was, in fact, becoming the most important educational institution of medical Melbourne; a position unimaginable in 1864 when the foundation Professor, George Britton Halford, conducted the first dissection class with just five others (including two medical practitioners who would later become his friends), in a shed in the backyard of his temporary residence in Madelaine (now known as Swanston) Street, Carlton. It was the future Chancellor and medical man, Anthony Collis Brownless, who had provided the energy and perseverance that allowed the School to be founded and to weather the early storms.[5] In fact, he started lobbying for its establishment from the moment he joined the University Council in June 1855, a few months after the inauguration ceremony. It was also his determination that meant that the duration of the course would be five years. The story of the rise of the Medical School from its infancy in the mid nineteenth century to its dominant place in Melbourne in the twentieth century, runs parallel with the meteoric rise of medicine as a science and the creation of a unified medical profession—one of numerous intriguing confluences of the history of medicine and the Melbourne School. As one of the most important drivers of this revolution, it is not surprising that the study of anatomy and its practitioners, the anatomists, were regularly at the centre of scandal and

George Britton Halford, Professor of Anatomy, Physiology and Pathology, 1862–82.
Courtesy: Medical History Museum, The University of Melbourne.

dispute. Notoriety was also guaranteed for anatomists, as anatomy was the subject of particular public interest in the nineteenth century.

Beginnings

The earliest classes in Anatomy in Melbourne at the University were held in 1855 at the commencement of first term. This was almost a decade before the founding of the Medical School. Among the subjects offered to first year Arts students was Comparative Anatomy, taught by the newly arrived Professor of Natural Science, Frederick McCoy (c.1823–1899). McCoy had studied medicine without taking out any qualification, and although his primary interests were in palaeontology, he would have certainly been a more than adequate anatomist. In fact, it is reported that in November 1839 at the age of just sixteen or seventeen, he twice lectured on anatomy at his father's school, the Peter Street Anatomical School.[6] Later, in 1848, he attended lectures at Cambridge in Medicine and Surgery without sitting the examinations.[7]

It was natural that comparative anatomy would be an important part of an Arts degree at that time. In the late eighteenth and nineteenth centuries a veritable flood of new flora and fauna found its way to Europe. This was a consequence of the race to colonise the unknown parts of the world. The appearance of so many new creatures meant that the study of comparative anatomy became central to understanding questions relating to creation itself. Australia was, of course, of great interest in this regard as it provided marsupials and creatures such as the platypus, to challenge orthodox views of biological classification. Comparative anatomy – especially in relation to marsupials – was to become a subject of particular and longstanding interest in the Anatomy Department at the University. In the eighteenth and early nineteenth centuries, comparative anatomists studied the different designs of living organisms in order to explain and catalogue all of God's creation as set out in the cosmological model of the Great Chain of Being, in which every creature was placed in a uniform ascending order from single cell creatures to God himself. Charles Darwin was to challenge this model in *The Origin of the Species* (1859), with the result that comparative anatomy was to play a central part in the battle between the creationists, led by Richard Owen, and the evolutionists, led by Thomas Henry Huxley. This debate was perhaps the greatest scientific intellectual debate of the nineteenth century. Owen was England's greatest comparative anatomist and, along with the eminent surgeon, Sir James Paget, was chosen to be the academic adviser for the appointment of the first medical professor at the University of Melbourne. The first anatomists at the University, McCoy and the first medical professor, George Halford, were to make sure that Melbourne was to play a small but significant part in the debate over evolution.

Considering McCoy's interest in anatomy it was natural that the first examinations conducted by the University would contain questions relating to human anatomy. Thus, question six in the Comparative Anatomy paper of 1856 asked 'By what characters could you distinguish a human blood stain, from one produced by the blood of birds or quadrupeds?'[8] Though little information remains of the subject in the first years of the University (other than the final examination papers from the surviving record of his lectures in 1872), it seems that McCoy taught the catastrophist theory of creation based on the work of the great French anatomist, Georges Cuvier, whose textbooks (along with Owen's) he used in the subject.[9] This theory posited a sequence of catastrophic extinction events with the subsequent creation of new life. It was accepted as orthodox, non-evolutionary thought as it allowed for God the creator, and explained the possibility of the existence of fossils of seemingly extinct species. In fact, McCoy had published a contentious monograph on extinct fossil fish before arriving in Melbourne. It is possible to gain an insight into McCoy's teaching of 'catastrophism' because the second professor in the Medical School, Harry Brookes Allen, preserved McCoy's lectures for posterity.[10] McCoy remained anti-Darwinian throughout his life, and it is only possible to understand why much of his Comparative Anatomy subject was primarily concerned with extinct marine fossils because he subscribed to the 'catastrophist' school. It seems that McCoy's anti-Darwinian stance made him a rather lonely and outdated figure in his latter years of lecturing, despite the important contribution he made to natural science in Australia. In 1898 McCoy's lectures in the early years of the medical degree were described in a rather unflattering light in a student journal. McCoy's students were, according to *Alma Mater*, connoisseurs 'in sponges, or plunged in ecstatic contemplation of extinct fishes.'[11] McCoy's subject Comparative Anatomy and Botany was removed from the medical course in 1881, to be replaced by Elementary Biology.[12]

On his arrival in 1862, Halford stood beside McCoy in the defence of Creationism against Darwinism and shared the same fate. This was partly because, as his successor in the Chair of Physiology and colleague, William Osborne, later put it, 'Halford worshipped Owen with the fervour of a disciple. He never abandoned this attitude and in later years was to suffer accordingly when the star of Huxley was in the ascendant.'[13] In the subject General Anatomy, Physiology and Pathology, Halford, like McCoy, used examples from comparative anatomy to raise objections to Darwin's theory, one of these being the best known of all—gaps in the fossil records.[14]

The teaching of anatomy in the University at the time was, however, no mere dry academic exercise, since anatomy was not only the cutting edge discipline used to expand the borders of medicine in a way never seen before, but it was also a topic of great fascination for the public. The intense intellectual interest in anatomy, specifically human anatomy, was stimulated not only by the intellectual pre-eminence of the study of comparative anatomy, but also by contemporary changes in the medical understanding of what constituted disease. This interest was mirrored in the public mind. Even before Darwin burst onto the colonial scene, the 1860s saw increased popular interest in both comparative anatomy and human anatomy. In his popular novel, *The Water Babies* (first published in 1863 and previously appearing in serial form in *Macmillan's Magazine*), Charles Kingsley frequently refers, somewhat ironically, to the anatomist as the popular representative of scientific wisdom. Writing of the possibility of catching a water baby, Kingsley suggests that the discoverer:

> Would put it into spirits, or into the *Illustrated News*, or perhaps cut it into two halves, poor dear little thing, and send one to Professor Owen, and one to Professor Huxley, to see what they would each say about it.[15]

In Melbourne, this interest in anatomy took the form of popular, privately-owned anatomy museums. An explanation for the popularity of such institutions is given in the catalogue of one of the first. This was Kreitmayer and Wiseheart's Grand Anatomical Museum ('Adults Only Admitted — ladies and gentlemen on alternate days'), which opened in 1861. The catalogue states:

> No subject has within the last few years attracted so much attention as the study of anatomy, and certainly none is so worthy of consideration when its bearing on the general welfare of the community is reflected upon ... Were the masses better acquainted with the formation of their own bodies than is at present the case, what maladies be prevented, what evils avoided.[16]

This museum was just one of at least four or five run by Maximillian Ludwig Kreitmayer, the last remaining a popular public entertainment until 1919.[17] Self-styled Professor Sohier, who eventually went into partnership with Kreitmayer, had opened the first general waxworks museum in Melbourne in 1857. Other than famous murderers, Sohier's exhibition specialised in phrenology, the popular practice of reading a person's personality from the bumps on the skull.[18] Phrenology had been invented in 1795 by the Austrian anatomist, Johann Franz Gall. Phrenologists believed that the brain was the organ of thought and that it was made up of a variety of sections that were related to different mental activities. By reading the shape of the outside of the skull (specifically the bumps) it was possible to understand the innate, biological mental abilities of the individual. It was not a big step to equate the differences in skull shapes to innate differences in the mental ability of various races. For example, in the *Phrenological Journal* of Great Britain, one could read that Hindus have small reflective powers and little combativeness; Papuan's skulls indicate thievery and a propensity to murder; and savages, apart from the love of their children, have no faculties higher than sensuality, cunning, covetousness, venery, cruelty and pride. Phrenology was popular in Australia and an antecedent of the craniometry studies of the third Professor of Anatomy, Richard Berry, at the University of Melbourne in the early twentieth century.[19] Cesare Lombroso took phrenology one step further and instituted the study of criminal head types, claiming that they came from a distinct group of degenerates that he called *homo criminalis*.[20] Richard Berry admired Lombroso. The Melbourne *Argus* newspaper stated on 9 November 1912:

> Some of us can remember when Lombroso was highly respected and his theories taken quite seriously. We all looked at his pictures of typical criminal heads, and were much impressed; and we were horrified when we noticed the 'stigma of criminality' in the heads of our personal friends.[21]

Phrenology became an immensely popular fad in Australia and, even though it was losing credibility by the 1840s, there were still many late in the nineteenth century under its sway. As late as 1898 the co-founder of the theory of evolution, Alfred Russel Wallace, came out in support of phrenology as a science that was able to define the 'mental peculiarities of the various races ... and accords with and explains national race and character'.[22] In 1894 Reverend Ralph Brown published *A Delineation of the Character, Talents, Physiological Developments and Natural Adaptations*, in which he discusses the animal propensities of both Ned and Dan Kelly.[23] Halford had a keen interest in craniometry and phrenology. He dissected the heads of the bushranger Morgan, as well as that of a Chinese individual and a number of Aboriginal heads, including that of King Jimmy of the Mordialloc tribe, all for the purposes of craniological comparison. He then supervised Richard Shepherd in their drawing and provided measurements ('suitable for the strictly scientific enquirer') for Brough Smyth's *The Aborigines of Victoria* (1878).[24] Such sensationalist material raised the profile of Anatomy in the public imagination. The criminal and indigenous population were also to be an important theme for many years and, indeed, throughout the tenures of the first five Professors of Anatomy.

Kreitmayer, the anatomy museum entrepreneur of Melbourne, was born in 1830 in Munich and abandoned medicine for anatomical modelling, supplying wax models to Sir James Paget, a key figure in the selection of the first professor in the new Medical School at the University of Melbourne. A trip to Australia and a stint in the goldfields saw Kreitmayer financially bereft and this led to him following his original profession of anatomical modelling in Melbourne. He was quick to see the financial benefits that could be gained from catering to the fad for anatomical information. Later, less scrupulous operators also sought to tap into this fascination leading to claims of chicanery and, finally, dramatic court action involving many of Melbourne's leading anatomists and surgeons. The 'clever feats of legerdemain ... by a professor of magic' that concluded evening medical lectures in these establishments fuelled accusations at the time that the motivations of the proprietors of such museums were 'extremely questionable'.[25] Kreitmayer seems to stand out from others who were less scrupulous, if only because he did not use the museum to bolster any medical practices.

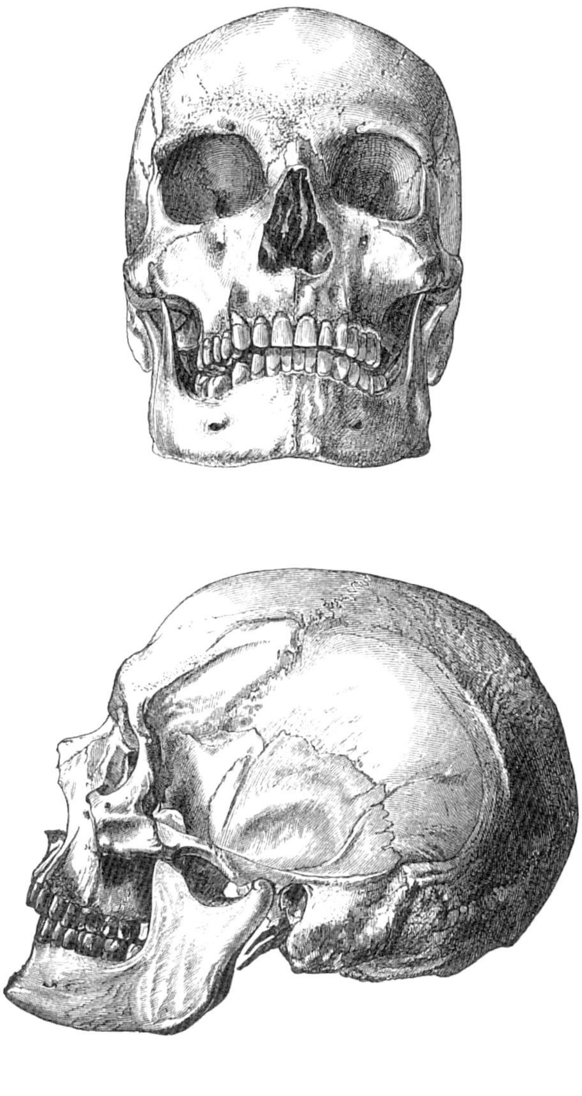

Skull of the bushranger 'Mad Dog Morgan' studied by Halford and illustrated in R. Brough Smyth's *The Aborigines of Victoria*, vol. 2, 1878, p.349.

Ned Kelly's death mask. Courtesy: Harry Brookes Allen Museum, The University of Melbourne. Photograph courtesy: Stuart Thyer.

Kreitmayer had a more direct connection, however, with the Anatomy Department at the University that relates to the bushranger, Ned Kelly. It is perhaps his only lasting legacy and dates from the 11 or 12 November 1880. At 10am on the 11th, thousands had assembled outside the Melbourne prison gates awaiting news of Ned Kelly's last moments. To satisfy the morbid interest of this crowd, within twenty-four hours after his execution, Kreitmayer had made a wax cast taken from the mould of Kelly's face and put it on public display in his waxworks museum in Bourke Street, Melbourne. Following the normal procedure after execution, Kelly's body was left hanging for half an hour before being taken down from the gallows. It was then carried in a handcart to the Melbourne Hospital morgue. Kreitmayer is understood to have then shaved Kelly's hair and beard, and made a mould of his head for the production of death masks before the body underwent a rigorous autopsy, finally being decapitated for more detailed phrenological study of the character of the bushranger.[26]

The teaching of anatomy, which was a topic of interest and importance in the public arena, had made a promising start in Melbourne under McCoy and even before the foundation of a medical school. The introduction of an undergraduate medical course in 1862 would see Anatomy eventually flourish as one of the pre-eminent subjects studied at the University of Melbourne in the following century, as well as being at the centre of some important debates in the broader area of national social policymaking.

George Britton Halford

On 23 December 1862, the first Professor of Anatomy, Physiology and Pathology, George Britton Halford, would have first noticed, when looking up from the deck of the ship as it sailed into view of Melbourne, a large building in a compound on the highest point in the suburb of North Melbourne. Early immigrants sailing up Port Phillip Bay were, according to folklore, 'dazzled by the sun shining on the open windows of the Benevolent Asylum, and wondered at the magnificent building shimmering before them at their long journey's end' as it was one of the first building sites that greeted 'the eye of the stranger as he sails into Hobson's Bay'.[27] Though increasingly hemmed in by surrounding buildings after its foundation in 1850, it still would have attracted the attention of those on ships until its demolition began in 1911. Subsequently, when discovering its purpose (to house those unable to care or provide for themselves), the new arrivals may well have wondered why a continent described by Charles Darwin as the 'great princess in the South'[28] had need for such a substantial home for the indigent. Victoria's gold heritage meant that old, poor, single men were overrepresented in the population at the end of the nineteenth century. And the itinerant nature of much of Victorian society, also caused by the movement between gold fields, added to a large part of the population barely able to subsist.[29] One consequence of this was that men outnumbered women by more than two to one in the Melbourne Benevolent Asylum in the late nineteenth century.[30] It has been said that, for many, this first view of Melbourne from the migrant ship, dominated by the Asylum, provided a warning of their future should failure attend their efforts in this land of opportunity.[31]

That the first landmark to be noticed by new arrivals was the Asylum was symbolic of life in the colony. Colonial Victoria, like many such societies, was a capricious mistress—frequently a place of boom or bust where riches seemed to force themselves upon some, while daily sustenance avoided others. This was certainly the case for the medical establishment. Many ships' surgeons and medically trained immigrants moved in and out of medical practice in Victoria and Australia generally as and when other opportunities to amass wealth beckoned.[32]

When Halford sailed up Hobson's Bay, the Asylum may have roused in him thoughts of even more concrete associations, because it was from such institutions in the United Kingdom that the material required for the anatomist's practice—cadavers for dissection—were obtained after the *Anatomy Act 1832*. He would have not then known that the Asylum and the antagonism of many in the medical profession towards the University would embroil him in scandal and difficulties in the years to come.

The Asylum in North Melbourne was a benevolent institution. Halford, on first experiencing the Melbourne medical community, must have wondered whether he was in Bedlam itself.[33] He told the Royal Society in 1867 that the endless feuds in the medical community in Melbourne needed to stop, so that medicine could be bought out of the 'mire in which it had been for so long, the medical societies being given over to backbiting and quarrelling'.[34] Halford was not, however, isolated in his new home, because despite the 'backbiting' nature of Melbourne medical society, many anatomists held important political roles in the early settlement of Victoria and the founding of the University. This group included not only Brownless, but also William Clark Haines, a surgeon before arriving in Australia, a member of the University Council and the first Premier of Victoria; Sir Francis Murphy, a member of The Royal College of Surgeons of England and the first Speaker of the Legislative Assembly of Victoria; and Sir James Palmer, Speaker of the Legislative Council and grand nephew of the painter, Sir Joshua Reynolds. Palmer was a qualified general practitioner and earlier in London had edited the four volumes of the anatomical works of John Hunter (1837). This was one of a number of connections between the Melbourne School and the Hunterian tradition.[35]

With his ability to dispense appointments at the University (though somewhat restricted by Brownless' control at the beginning), Halford was in a position to make powerful friends and equally as powerful enemies. The potential for trouble was further increased because Halford was an excitable individual with a volatile nature, and placed in the Melbourne medical community, the possibility for his involvement in debate and schism was real. Almost as soon as he arrived, jealousies were in evidence.

One of his early contacts was with David Thomas, surgeon at the Melbourne Hospital, and the first examiner in Anatomy at the University. Thomas had taken himself on a European tour between 1853–9 where he obtained his doctorate at St Andrews University, a Fellowship from The Royal College of Surgeons of England and studied anatomy and histology in Paris. In the winter of 1858–9, Thomas studied for four months at Heidelberg under Professor Nuhn (after whom some of the glands of the tip of the tongue are named). His testimonials from these studies highly recommended him as a teacher of anatomy.[36] To Thomas' disgust, upon arrival, Halford had gained the permission of the University Council, after much persuasion, to conduct a limited private medical practice. The Council only allowed Halford to practice at the request of, and in the presence of, another medical practitioner.[37] In a highly competitive medical marketplace, Thomas believed that such activities were not in keeping with Halford's position as the Professor at the University.[38] It seems also that Thomas may well have wished the Anatomy Chair for himself. When he applied for the lectureship of surgery in 1864, he demanded the position be upgraded to that of a professor along with a commensurate salary of £500 per annum, half that of Halford's, but £400 more than that paid to the lecturers in Law.[39] Halford would have presented, of course, considerable competition for other anatomists such as Thomas. Thomas resigned as Examiner in Anatomy after two years of working with Halford.[40]

In his early years in the Chair, Halford had many other reasons to be anxious and problems beset him on all sides. The number of students in the course rose little from the first intake of three. By 1872, there were only thirteen first year students—hardly a ringing endorsement of the School by Melbourne's middle class. The first cohort graduated in 1867. In 1868 only two students graduated (excluding honorary degrees); in 1869, only one; and in 1870 and 1871, none at all.[41] In 1871, nine years after the founding of the School, there were only five medical practitioners in Victoria who trained in Melbourne, as opposed to four hundred and three trained in Britain. In 1881 there was still only thirty-four as opposed to three hundred and eighty-nine.[42] Melbourne medicine remained very much under the sway of British doctors. The British *Medical Examiner* reported in 1877 that:

> An English surgeon, on landing in Melbourne will find all matters connected with his profession very much like those at home. Walking up Collins Street East, he will see the brass plates of doctors as thick as in Saville-Row or Wimpole-Street: on one he will see written "Victorian Medical Association", on another "Victorian Medical Benefit Association"; and opposite many of these brass plates, carriages, with horses well-groomed and coachmen well-dressed, proving the existence of handsome incomes. Elsewhere our stranger will see two large general hospitals, a children's [hospital], a lying-in, an eye and ear, an orthopaedic, and other similar institutions.[43]

Many prospective students, including the sons of some members of the University Council,[44] continued to sail to England and Scotland for the shorter four-year course that contained much more extensive clinical practice.[45] Further discouragement presented itself to prospective students due to the difficulty of completing the course. Failure rates at the Melbourne School among students were always near the fifty per cent mark.[46]

Amidst all these problems, and notwithstanding he was considered by Paget and Owen to be 'one of the most distinguished experimental physiologists of his day',[47] Halford must have soon realised that his

Halford (second from right) and students dissecting in 1864.
Courtesy: Medical History Museum, The University of Melbourne.

opportunities for research in the colonial outpost would be overwhelmed by the strenuous efforts needed to set up a Medical School, teach a large proportion of the curriculum and make some extra money in medical practice, so as to feed a rapidly growing family. To make matters worse, in his first few years after arriving, he found himself embroiled in numerous controversies. In July 1863, he was caught up in a dispute with 'Darwin's Bulldog' Thomas Henry Huxley, over Darwin's evolutionary theory and in 1864, with the redoubtable Sir Redmond Barry, Chancellor of the University. In the latter case, this was all because of Halford's interest in phrenology and other related sciences of the brain. In the first instance, as we shall see, he irretrievably damaged his chance of preferment in Britain, since those sympathetic to Darwinism increasingly occupied all the major decision-making posts in Britain's scientific establishment in the second half of the century. Thus, Halford's attempts for recognition in the home country, such as a fellowship of the Royal Society of London, met with failure. His unsuccessful efforts to find an effective chemical treatment with ammonia for snakebite would have not aided his cause.[48] Moreover, in the controversy with Barry, he tarnished his reputation with the Council of the University—his sponsors in Melbourne.[49]

The 1864 conflict with Barry and the Council concerned the case of the murderer Harrison, at the time known as 'the William Street tragedy'. It was a cause célèbre, and widely reported in the press, both in Melbourne and elsewhere. On Saturday 18 June 1864, outside the office of the Roads and Bridges Department in William Street, Christopher Harrison shot his partner James Marsh over a financial dispute. He then attempted to commit suicide but the pistol failed to discharge.[50] The London *Times* reported that rumours of insanity soon started circulating, especially due to the cold-blooded nature of the act and Harrison's attempted suicide. And Harrison's son boosted such stories by claiming that there was insanity in the family in England.[51] The Harrisons' grocer's assistant claimed that upon visiting Harrison he would enthusiastically begin discussing 'geometry, trigonometry, algebra, bases, perpendiculars, hypotenuses etc', but that his manner of speaking never had 'any proper coherence or suitable connexion [sic] between the odd ideas which were jumbled out'.[52] He apparently also frequently held his temple during these ramblings—evidence of a pathological brain. It was said that Harrison had suffered an injury to the skull at some previous time. Amid the clamour for a finding of insanity and reprieve from hanging, Halford was the most persistent in claiming Harrison's innocence.

As the former Medical Superintendent of the Liverpool Lunatic Asylum, Halford claimed special expertise in recognising Harrison's madness. He claimed that the evidence pointed to Harrison's behaviour as the product of lesions on the brain; perhaps the first example in Australia (and one of the earliest in the world) of the new diagnostic methods that were being developed in medicine and applied to the brain. It had only been three years earlier, in 1861, that Paul Broca conducted a post-mortem on an individual named Tan, who had suffered from aphasia (an inability to understand speech or otherwise to have some form of speech loss), in which he demonstrated the existence of a superficial neurosyphilitic lesion in one side of the brain, precisely in the area which Broca believed controlled speech. This area became known as Broca's lobe. It was not until 1876 at the earliest, that the Scottish surgeon, William Macewen, performed an operation to remove an abscess from Broca's lobe in a patient suffering transient aphasia.[53] Halford's diagnosis of insanity due to lesions on the brain was, therefore, a very early example of the application of neuropsychology. 'Hang such a man in Melbourne in 1864! Oh, fie!' was Halford's comment in the *Argus*.[54]

Halford was much ridiculed in the popular press for his efforts.[55] The *Leader* expressed the views of Melbourne journalists when it commented in its editorial on 6 August:

> It is a lamentable spectacle to see a man, supposed to be possessed of superior attainments and occupying a position of some importance in society, so entirely blinded of self-conceit as to forget what should be the behaviour of a gentleman, and, consequently, degrade himself in the eyes of the public. This spectacle was presented by Professor Halford, at the meeting held on last Monday, at Harwood's library, for the purpose of taking in to consideration the question of Harrison's sanity, and memorialising the Governor for a respite.[56]

And what was the reason for such spite? As his first act as chairman of this public meeting, Halford had attempted to eject all the representatives of the press because he alleged:

> after a long experience, he had ascertained that these gentlemen, either through incapacity, ignorance or wilfulness, invariably reported partially and incorrectly any remarks involving matter of scientific interest.[57]

Halford had managed with one intemperate remark to make himself a prominent target for public criticism. What is more, the government took no notice of him at all. The 'red-handed' government refused to be moved. Harrison's fate was sealed and on 3 August 1864 he was hanged, along with two bank robbers, Woods and Carver. On the scaffold, Harrison bequeathed Halford his brain for dissection in recognition of Halford's attempts to save him, and in order that an autopsy could further the study of neurology. What followed next was reported at length in all Melbourne newspapers.[58]

The Lecturer in Surgery at the University, Edward Barker, customarily performed the post-mortem examinations after executions. As such, the *Illustrated Melbourne Post* informed its readers, Barker had the heads of the executed criminals conveyed to the University because Dr Macadam, the lecturer in Chemistry, and an early exponent of forensic medicine, was intending to ascertain the specific gravity of each brain 'and otherwise experimentalise [sic] upon them'.[59] Halford, however, arrived at the University before the others and commenced a dissection of the skulls and brains, despite earlier evincing no interest in them. This performance in the Anatomy theatre at 3pm was before 'a brilliant array of medical and surgical talent'—around thirty or forty in number.[60] Having removed the brain from the skull, Halford, so the *Illustrated Melbourne Post* reported:

> handed around Harrison's brain, and pointed out a depression on the right hemisphere, with a corresponding depression on the frontal bone. A plate, containing a small portion of substance, which was stated to be a whitening of the arachnoid membrane, from under the dura mater, was then submitted for examination.[61]

Halford had his evidence. Even with such an incomplete and vague description, it would be possible to surmise now that such an injury to the right lobe could result in impairment of Harrison's higher intellectual functions.

At this moment 'a side door opened' and in walked Dr Barker supported by a 'formidable Sheriff, 6ft 4in in height'[62] along with a justice of the peace, a policeman and Macadam. The *Post* described the chaotic scene to its readers:

> Dr Barker excitedly complained of the professor's conduct in walking away with the brains ... Dr Halford said that they had been sent to him, which was most emphatically denied by the Surgery Lecturer. Professor Halford asserted his rights to the remains, as he had seen by the papers that Harrison had left them to him. The Sheriff said that the brains were his property, as representing the Crown, and a convict could not dispose of his property ... Professor Halford then stood on his dignity, and asserted that he was the only licensed anatomist in the colony, and the most competent person to conduct the examination. He would not be interfered with in his own theatre

> in his own hours, but that when he was finished Dr Barker might do as he pleased. He said that had Dr Macadam steeped the brains in any solution, they would afterwards have been useless for his purpose.[63]

Whether Halford's stubbornness or his arguments prevailed, Barker finally acquiesced to the plan that the dissection would go ahead, and the sheriff agreed. All three brains were dissected and analysed in front of all present during the dispute. It is difficult to know whether the interruption had unnerved Halford, but he concluded that Harrison's brain was not pathological, but quite healthy. Always controversial and pigheaded, at the very moment he should have been willing to be brave and back his early proposition, he hesitated and retreated. And Halford missed his moment.

A correspondent to the *Argus* described the farcical scene under the title 'The battle of the brains':

> The tenacity with which Professor Halford has clung to his subject, living and dead, is a spectacle in the last degree touching. The story over the battle of the brains presents a peculiar mixture of the horrible and the ludicrous, and forms a fitting climax to the latest instance of the madness of the mad doctors ... Over the ghastly heap of dead man's brains, rages fierce war between the rival anatomists.[64]

Halford's stocks had fallen. Only one year earlier, many newspapers had supported Halford strongly in his attack on Huxley and Darwin. Perhaps *Melbourne Punch* best summed up the sentiments of the press in its satirical poem *A Doctor's Difference*:

> Was Halford right or Halford wrong?
> Was Barker wrong or was Barker right?
> The error may to both belong,
> But t'was a most unseemly fight.
>
> The trembling limbs had scarce grown cold,
> The vital spirit scarce had fled;
> Ere learned men each other scold,
> And loudly clamour for the dead.

One doctor cries, "The brains are mine,"
And stands upon his dignity;
We trust his audience *could* divine
How firm a footing that must be.

The other answers, "You mistake,
The brains belong unto the Crown;
Yet for our learned brethren's sake
My friend and I will knuckle down."

We deem it not our place to show
Which side had reason in the fight;
But *should* have brains enough to know
Such disputations are not right.[65]

To add insult to injury for his role in the Harrison matter, the University Council censured Halford. The Vice-Chancellor, Sir Redmond Barry, was furious.[66] So much controversy at such an early stage in his occupancy of the Chair harmed Halford's reputation. An unnamed ex-president of the Medical Society of Victoria told *Argus* readers in 1867:

> Professor Halford has neither ... candour [nor] ...knowledge of the world. He is a good anatomist and an excellent man, but, unfortunately for his reputation and influence, he has some grievous infirmities ... he is rash and impulsive, and blurts out whatever he happens to think at the moment, without the least regard to the views and feelings of others ... whenever he fails to get his own way, or meets with any opposition to his views, he takes it at once as a personal offence, flies into a pet, and fancies the whole world is up in arms against him.[67]

Such was Halford's fall from grace that on 28 June 1869, McCoy was allowed by the University Council to give a public lecture in the Princess Theatre attacking Huxley's protoplasmic theory (that all life can be described materially without the need for the divine). Halford's request at the same time for Council permission to deliver a public lecture was refused. As we will see, other factors were also at work.[68]

It also seems that the end of his career was less than happy. His erratic behaviour in the 1890s, including excessive drinking, was of ongoing concern to the Council and the School.[69] Speaking of his later years, W A Osborne, a successor as Professor of Physiology, relates that: 'a generation of Melbourne doctors has memories chiefly of a man weary of his task'.[70] And to make matters worse, while his early work on the physiology of the heart had been groundbreaking, he seemed increasingly aware that he had wasted much of his research time on a subject (namely ammonia as an antidote for snakebite) that was an intellectual cul-de-sac. His retirement in 1896 ended the connection with the beginnings of the Medical School. As an eminent surgeon, Alan Newton, later remarked, 'Halford made Melbourne and Melbourne marred Halford'.[71]

Melbourne's medical marketplace

The medical profession in Melbourne in the mid nineteenth century was truly a 'medical marketplace'. As the leading article of the *Leader* newspaper claimed on 28 August 1869, the Medical Board was compelled to admit 'anyone who, by examination or purchase, has obtained some sort of professional degree from any "body duly recognised in the country to which such body may belong"'.[72] This meant that the medical profession had no unifying qualification. Apothecaries, surgeons and physicians (all with a bewildering variety of qualifications) vied for patients and their money. For example, in 1881, those who were allowed under the *Medical Practitioner's Act 1862* to treat the sick for a fee included not only four hundred and fifty four registered medical practitioners, but also almost one thousand assorted alternative medical practitioners, such as apothecaries, midwives, dentists, Chinese doctors, homeopaths, galvanists, mesmerists and hydropathists, among others.[73] All this explains the regular fracas in Melbourne over the propriety of various schemes to attract patients. Rumours concerning the legality and even veracity of many of the qualifications claimed by medical practitioners were the bread and butter of medical Melbourne at this time.

The leader of the raffish clique of Melbourne surgeons, the irrepressible James George Beaney, is an example of such gossip.[74] Beaney or 'jelly-belly' as he was called in the *Melbourne Medical Record*, was one of

Melbourne's most successful surgeons and, according to his enemies, a plagiarist, fraudster and consummate self-promoter, as well as undeniably ostentatious and a man who enjoyed the high life. Starting his career as an apothecary's assistant in a Collins Street chemist shop, he took himself to Edinburgh and obtained a diploma at The Royal College of Surgeons of Edinburgh in 1855 during the Crimean War at a time (according to his critics) when the acute shortage of surgeons resulted in a significant dropping of the standard needed to pass.[75] His public extravagance, funded by his tremendous success, antagonised those who despised his methods, or were jealous of his achievements.

Dr G T Howard, one of Beaney's residents at the Melbourne hospital, reported Beaney expressed much surprise:

> When a patient rejected his offer of a glass of champagne in favour of a cup of tea, "Augh, tea", he snorted and then jauntily to us, "I am becoming quite continental in my habits. I take a bottle of claret for breakfast, a small bottle of fizz for lunch and a big bottle for dinner", and he gently stroked a very obvious protuberance.[76]

In his typically ironic, cheeky style, Beaney summed up the medical scene into which the Medical School was born and into which Halford arrived. He told an audience in Melbourne in 1876 that while discord ruled in medical communities the world over:

> In Victoria ... "Never, Never, Never!!!" We are the best of friends, and the closest of cronies. We never disagree as to our methods of treatment, surgical or therapeutic. We never write anonymous letters to the newspapers, for the purpose of decrying the skill, or injuring the reputation of a brother practitioner. We never make mistakes, such for example, as delivering a child, and forgetting to bring forth the infant's head with the rest of the body; never operate for "phantom tumours", never remove the entire bladder while professing to extract a tumour only. Never tell a man ... with a compound fracture that it is nothing but an ordinary bruise, and send him [away] so that the gangrene sets in, and the poor fellow's leg has to be amputated; never pronounce a broken back to be a mere strain; never perform surgical operations while under the influence of liquor ... In fine, whatever dissensions may prevail amongst journalists – for even they do not always

Melbourne surgeon James George (Diamond Jim) Beaney.
Courtesy: Medical History Museum, The University of Melbourne.

> agree – or among clergymen, musicians, actors, and actresses, public singers, lawyers, merchants, politicians, and so forth – who is acquainted with the medical faculty in Melbourne would venture to insinuate that Doctors differ?[77]

Beaney, although occasionally successful in the elections at the Melbourne Hospital (notably in 1875 and as a result of bribery, so claimed his opponents) never shared the same success in gaining a position at the University.[78] Halford was one of those who had little time for Beaney. Indeed, Halford was among a group of medical men who acted as expert witnesses in malpractice cases against him. Perhaps the most famous case that involved Halford giving evidence against Beaney was the death of Mary Lewis in 1866 after a botched abortion.[79] The others in the Halford camp were William Pugh, the first surgeon to administer an anaesthetic in Australia; James Thomas Rudall, honorary surgeon at the Melbourne and Alfred Hospitals, ophthalmologist, examiner in Pathology and Physiology at the University and the first examiner of Anatomy, along with David Thomas, in 1862;[80] James Edward Neild, lecturer in Forensic Medicine at the University, later forensic pathologist for the State, editor of the *Australian Medical Journal* between 1862–79 and better known as one of Melbourne's leading literary critics; and Richard Thomas Tracy, lecturer in Obstetrics, regarded as Melbourne's most distinguished obstetrician.[81] All of them had violent disputes with Beaney at some time or other, and these men were to become important supporters of Halford and the Medical School. It is possible to see in these events the beginnings of the unification of the profession under the umbrella of the nascent School, even though Halford's team was unsuccessful and Beaney won his cases, using the notoriety so gained for the purposes of self-promotion.

Inclusion in the inner sanctum of the School precinct was not just refused to the underqualified entrepreneur. Intellectual nonconformity could also tarnish a reputation. The most active proponent of contagion and germ theory and Darwinism in Melbourne, William Thomson, was denied appointment even when he was the only candidate for a position.[82]

The case of Edward Barker (after whom Barkers Road, Kew is named) is a good example of the way in which loyalties ebbed and flowed in this community. Barker and Beaney were great enemies. They regularly appeared as expert witnesses in cases where the other was being charged with surgical malpractice—a not uncommon charge at the time. Initially, this was to Barker's benefit. Barker's preferment illustrates the early desire by the University to avoid scandalous behaviour on the part of its staff. As Selleck has argued, the University Council required that prevailing orthodoxy not be disturbed by members of the University.[83] This could be difficult at times as Halford proved. As the first Lecturer in Surgery at the University, Barker held an important place in the development of a unifying medical accreditation system that the University would eventually impose over the diverse Melbourne medical marketplace. Respectability, both intellectual and social, was important in ensuring the victory of the University. Twice, however, he offended sensibilities. Russell tells the story of the Council's refusal to reappoint Barker in 1873 and its subsequent retreat. He remarks that the decision not to reappoint Barker was instigated by Brownless, but fails to mention the damaging and widely discussed law suits that Barker had previously fought over surgical malpractice, as well as the charge relating to an accusation of impropriety by a Mrs Tiller in 1869. She accused him of attempting repeatedly to kiss her in his rooms. From the reports of the hearings concerning this episode, it seems that Barker was well served by his professional colleagues on the investigating board.[84] None of this would have pleased Brownless, a devout Roman Catholic. Then, in 1880, the University Council received a complaint from medical student, Richard Richards Harvey, claiming that Barker was intoxicated while delivering a lecture. Although the Council cleared Barker, he was not reappointed to the teaching staff.[85] Barker's replacement was Tharp Mountain Girdlestone, a leader in the introduction of antiseptic surgery and possibly the best surgeon in Melbourne at the time.[86] Girdlestone had also supported Halford, and his reputation as a surgeon qualified him eminently for the position—an example of the changing alliances of the time. A personality such as Beaney could not be completely expunged, however, by the normalisation of the profession. His generous endowment for scholarships has benefited the very School that was to make practitioners such as Beaney obsolete.

A dispute between Beaney and Barker concerning an anatomy museum also highlights the contemporary public interest in anatomy. Furthermore, the dispute illustrates the volatile nature of Melbourne's medical establishment. The cause of all the trouble was the celebrated court case concerning the Anthropological Museum (for gentlemen only) set up in

September 1867 at 172 Bourke Street East by Dr Jordan and Dr Beck. The museum contained the usual collection of wax models of female, male and anatomical parts, including those of numerous pathological organs illustrating exotic and rare conditions. The cost of admission included a copy of *The Philosophy of Marriage* by Jordan and Beck, two catalogues and a pamphlet. Included in the material was an advertisement for their medical practices in Latrobe Street, Melbourne and Albert Street, Ballarat.[87] On Saturday, 4 September 1869, the *Age* newspaper lashed out, claiming that originally waxworks anatomy museums had been set up for laudatory reasons:

> For the most part the models exhibited were really instructive and were characterised by some artistic merit. The approval of the first visitors having been obtained, the proprietors gradually developed what appears to have been their ulterior plans. They accumulated models and castes of the filthiest kinds until the chief feature of the show was the horrible pathological distortions which the instincts of mankind have ever kept as secret as possible.[88]

The *Age* argued that such sensationalist exhibitions were used to direct the uneasy visitors to the proprietors' medical rooms:

> The principals make the museum subservient to a more profitable shop which they keep elsewhere ... these doctors practice in what medical men look upon as the lowest grade of the profession. But it is certainly not the least lucrative, as patients find out to their cost.[89]

Dr Jordan sued for libel and the proprietor of the *Age*, David Syme, briefed George Higinbotham, later to become Chief Justice of the Supreme Court. The case reveals the disreputable side of these anatomical museums. This particular one was designed to panic its customers into attending the medical practice, run by unqualified practitioners. Syme won, resulting in the closure of that particular museum. What was rather extraordinary was that Edward Barker gave evidence for Jordan, claiming that he:

> found no fault with the museum on his brief inspection before its opening ... He thought inspection of the museum might perhaps do his son (then aged 15) good in a few years' time ... It might depend on taste whether the exhibition was considered pleasing (Laughter).[90]

The scandal of one of Melbourne's leading establishment surgeons supporting such an enterprise led to the convening of a special meeting of the Medical Society of Victoria, of which Barker was a prominent member. The reason given for calling the meeting was the 'reprehensible conduct of one so esteemed and eminent, giving evidence favourable to a notorious quack and an exhibition necessarily repudiated by all respectable medical men.'[91] Barker was forced to eat humble pie, both apologising to the Society as well as producing written evidence that he had returned a donation from Jordan and associates of £21 to a benefit for himself! James Beaney was overheard to remark on the honesty and motive of Barker that:

> When the Anthropological Museum was established invitations were issued to the profession to inspect it. He did not go, but he believed Dr Barker did ... He had it from the gentlemen of the press, upon whose veracity he could rely, that when Dr Barker saw the museum he stamped his foot upon the ground and said it was a beastly and disgusting exhibition. Why he had changed his tone since that time he was at a loss to understand, and so were many others.[92]

This episode was merely one of many similar such upheavals, including the ongoing competition for the elected honorary positions at the Melbourne Hospital and various positions at the newly founded Medical School. Halford was to be involved in even more controversy – this time in his attempts to find enough cadavers to teach Anatomy in the new School. This story is told in Chapter Four.

The subject of anatomy had been the cause of the most extraordinary public interest and controversy within nineteenth century Melbourne medical circles and in the wider community. The first Professor of Anatomy, George Halford, was gifted with ability, but was hampered in his role as leader of the new School due to his excitable and forceful character, as well as with causes beyond his control. He was ill suited to pour oil on the troubled waters of Melbourne's medical scene. It was unfortunate for Halford that the first three decades of the teaching of anatomy had been beset with the most extraordinary series of debates and scandals. His student and successor was to provide more sober direction for the growing Medical School. The last two decades of the nineteenth century were to establish the reputation of the School and the study of anatomy as the bedrock on which that reputation was built.

CHAPTER TWO

The professors versus Darwin

THIRTEEN YEARS AFTER the Duke of York attended the riotous Commencement Ceremony, on the evening of Wednesday April 29 1914 the Medical School celebrated its coming of age at the celebrations for its fiftieth anniversary. This took the form of a sumptuous dinner for one hundred and eighty medical graduates of the University of Melbourne at the Grand Hotel (now known as the Windsor). Such an occasion at one of Melbourne's finest and grandest dining establishments would have removed any lingering doubt that the Faculty, as the School had become, had failed to cast off any sense of inferiority that may have characterised the early years. Not that confidence was in short supply among the earliest graduates. As early as 1877, an English observer had noted of Melbourne medical students that: 'these graduates have an overwhelming belief in their own powers, think the questions set by their examiners the most difficult in the world, and generally have a great idea of their degree and attainments'.[1]

A bevy of political and university dignitaries also attended the celebration. In recognition of the status of the Faculty in Melbourne, the chairman of the proceedings was Sir Arthur Stanley, Governor of Victoria. The celebration of the achievement of reaching the respectable age of fifty years was further enhanced for the University's medical establishment because Melbourne's Medical School was the first English-speaking School to be established in the southern hemisphere. It was also the first colonial School to have its degrees recognised by the General Medical Council of Great Britain in 1890. During the course of the evening, George Halford's student and successor, Sir Harry Brookes Allen, Professor of Pathology and Dean of the Medical Faculty (as the School was then named), replied to the toast given by Mr John Grice, President of the Melbourne Hospital. His address was not merely a recital of gilded successes, although initially and in keeping with the ebullient atmosphere, Allen spoke of the tremendous achievement of fifty years of medical education in Melbourne. He soon assumed a more sombre tone, however, and conceded that the pace of revolution in science and medicine had thrown up serious challenges for staff and students. He spoke of the great difficulties of the early years of the School and the labours that allowed it to survive and then flourish. What made establishing medical education in Melbourne even more difficult were, he explained, 'the tremendous changes which have taken place in scientific methods in the first fifty years of the Medical School', changes which the School did not always accommodate easily. Allen listed those changes for his audience. First there was the discovery of the periodic law in chemistry described as the 'Alphabet of Creation', and the 'coming of the doctrine of Evolution, which has transformed our whole outlook on the world around us, and on ourselves even in our highest relation'. Also in this period appeared 'antiseptic surgery which ... revolutionised all surgery' along with the beginning of the new science of bacteriology. In professorial fashion, Allen instructed his audience to 'think of the great men who have been responsible for these new developments – Mendeleef, Darwin, Lister, Pasteur, and their successors – and you will see how all opinion has been in a state of flux, and the task of teachers and taught [students] has been correspondingly difficult'.[2]

Charles Darwin's revolution

The Medical School had indeed initially struggled to keep up with the pace of the discovery of new ideas. Throughout its first fifty years, however, the study of anatomy had been one constant in that sea of change. The assimilation of Charles Darwin's theory of evolution as found in the *Origin of the Species*, is one telling example of the sometimes reluctant acceptance of new ideas in the early decades of the School. In this case, the study of anatomy at the University was directly implicated in the tale.

The appearance of Charles Darwin's *Origin* has been characterised as the symbolic moment of change from the old, pre-modern cosmology of a static creation, to the modern world of evolution, change and scientific progress. The notion of a permanent, unchanging creation was explained in the work of Bishop James Ussher in the seventeenth century. In *The Annals of the World*, he calculated that the cosmos was created on Sunday 26 October 4004 *BC*, Adam and Eve were driven from Paradise on Monday 10 November 4004 *BC* and Noah's Ark touched Mt Ararat on Wednesday 5 May 1491 *BC*. In the late eighteenth century the Scottish geologist, James Hutton (1726–97), raised the idea of an extremely ancient earth ('No vestige of a beginning, no prospect of an end') developing the earlier work of Benoît de Maillet (1656–1738). The concept of deep time had been created and evolutionary theory had its essential precondition. Darwin was not the first to postulate the idea of evolution—nor was his book the most popular text. Robert Chambers' *Vestiges of the Natural History of Creation* (1844) far outsold Darwin's. Although Darwin's theories were just part of a long gestation of evolutionary thought, for many, it was Darwin who had delivered the final blow to the idea of the hand of God in creation. The *Origin of the Species* was available for sale in Australia five months after its appearance in Britain and became a hot topic of discussion among local intellectuals.[3] The appearance of *Origin* and Thomas Henry Huxley's *Man's Place in Nature* in 1863 in the period covering the foundation of the School, are two confluences that characterised and formed the history of medical, particularly anatomical education in Melbourne.

Before 1859 scientific creationists held virtually all the important scientific positions in the English-speaking world. At first, Melbourne's intellectual circles held views similar to those of the English critics and a majority rejected Darwin's version of evolutionary theory, especially natural selection. From the beginning, and for many years, the medical and biological arms of the University resisted the tide of evolutionism through the efforts of Professors McCoy and Halford. William Hearn, the foundation Professor of Modern History and Literature, Political

Economy and Logic, was the lone Darwinian voice in the early years of the University, publishing a defence in 1863.[4] In Melbourne medical circles, it would be reasonable to say that acceptance of Darwin's theory grew more quickly outside the University than from within.

The battle between the University anti-Darwinians and their opponents is one of the most interesting episodes in the history of the School and of the teaching of anatomy in Melbourne. The most audacious episode in this saga was Halford's attack in 1863, soon after his arrival, on Thomas Henry Huxley. It was anatomy that was central to the dispute as the contested ground concerned comparative anatomy.

By the 1880s Darwin's opponents in the United Kingdom had lost control of most of the senior scientific positions and evolution had become the dominant creed in intellectual circles, though not natural selection.[5] The only significant exception to the victory of the Darwinists in the English-speaking world in filling the upper echelons of the scientific hierarchy after this date were the long reigns of the catastrophist (and the discoverer of Ice Ages) Louis Agassiz and his student, Nathaniel Shaler at Harvard.[6] It should be remembered, however, that an important reason for the acceptance of Darwin's theory of evolution was that it became incorporated into the Victorian paradigm of 'progress'. This version of evolutionary thought was seen to have direction or a hierarchy and goal, as, for example, in the superiority of the white races over their coloured cousins. According to the protagonists, it was not random selection that saw the whites on top; rather it was a product of an inherent superiority of all the important human characteristics. Self-help was also central to the professed superiority of the white races, particularly the male sex. Lamarckianism (a late eighteenth century French version of evolution) again became popular because it emphasised the importance of the agency of the individual. Improvements gained by parents could be passed down the generations. The classic example given at the time was the blacksmith's son putatively inheriting bigger biceps. There was no place for the blind chance of pure natural selection, a Lamarckian position which Darwin always included in his writings alongside natural selection from the first edition of *Origin*.

It is also important to emphasise that Darwin's first opponents were not simply a group of rigid creationists. Originally, opposition to Darwin included a wide spectrum of religious and scientific thinkers. At one end of the spectrum were those such as Bishop Samuel Wilberforce, Huxley's famous opponent in the much celebrated and misreported debate at the Oxford meeting of the British Association for the Advancement of Science in the new Oxford Museum in 1860.[7] Wilberforce argued for the biblical version of creation, but many other opponents of Darwinism accepted evolution generally but disliked the idea of natural selection, that is, the random choosing of individuals for survival and procreation that seemingly removed God from the act of creation. Thus, they were not prepared to rule out some form of direction and they formulated theories of evolution along those lines. Most nineteenth century Darwinists eventually followed this path as it fitted Victorian ideas of progress—evolution was a steady ascent that only the best could fully undertake. Other than strict creationists, theistic evolutionists such as Richard Owen (the first to coin the term 'dinosaur' and who had classified many of Darwin's fossils from the Beagle), believed that variation was an active force driving species in a predetermined way. This could allow for a role for the supernatural, which meant that it was necessary to retain the authority of religious leaders who would interpret the moral significance of the creator's activity.

Darwin in Melbourne

What were the consequences in Melbourne? Halford's stoush with Huxley originated in a dispute between his mentor Owen and Huxley. As discussed earlier, Professor Owen had chosen Halford for the foundation Chair in Melbourne. Owen, the pre-eminent comparative anatomist in England, classified many of the new discoveries of species, including in 1847 a newly discovered species of ape, the gorilla. This discovery stimulated the debate about apes, evolution, mankind and the connections between them. The University of Melbourne Medical School was drawn into the debate over man's place in nature, when Owen defended the orthodox view of mankind's origins, by questioning the notion that mankind was descended from the apes, or at least shared a common ancestor. Huxley proposed the evolutionary view, most famously, in the 1860 debate in the Museum in Oxford against Bishop Samuel Wilberforce.

Halford took up the defence of his mentor Owen whose anti-Darwinian views led him to state that gorillas and other apes lacked a structure in their brain that is found in humans, namely, the *hippocampus minor*, thus showing that humans could not have evolved from apes. To the satisfaction of the scientific community Huxley proved they did not lack the organ, which tarnished Owen's reputation greatly. It seems, however, that the method of preparation used to preserve Owen's gorilla brain may have caused the destruction of the *hippocampus minor*. Charles Kingsley satirised the debate in his children's classic, *The Water Babies*:

> You may think that there are other more important differences between you and an ape, such as being able to speak, and make machines, and know right from wrong, and say your prayers, and other little matters of that kind; but that is a child's fancy, my dear. Nothing is to be depended on but the great hippopotamus test. If you have a hippopotamus major in your brain, you are no ape, though you had four hands, no feet, and were more apish than the apes of all aperies. But if a hippopotamus major is ever discovered in one single ape's brain, nothing will save your great- great- great- great- great- great- great- great- great- great- great- greater- greatest grandmother from having been an ape too.[8]

As we have seen, McCoy, Halford's co-creationist, was probably a catastrophist like Agassiz. It was McCoy's efforts that established the National Museum as an important cultural institution in Melbourne. All the displays in the Museum, which had been relocated to the University in 1856 and housed in an impressive new building from 1863, were arranged to prove creationism and discredit Darwinism.[9] The centre point was a large display with one of the earliest collections of gorillas in the world, purchased by McCoy in 1865 from the explorer who first exposed gorillas to the attention of Europe, Paul du Chaillu.[10] As the *Argus* proclaimed:

> it is well for the inhabitants of a country so remote, under ordinary circumstances, from the chance of seeing actual specimens of this the greatest and most man-like of the anthropomorphous apes, to see how infinitely remote the creature is from humanity, and how monstrously writers have exaggerated the points of resemblance when endeavouring to show that man is only one phase of the gradual transmutation of animals.[11]

McCoy's gorillas, used by Halford as illustrative material in anti-Huxley lectures at the Royal Society of Victoria, were to become unwitting players in the dispute that consumed scientific and medical Melbourne.[12]

Halford versus Huxley

At first Halford seemed to reject Darwinism with a religious fervour, especially in relation to the place of humanity in evolution. On 17 July 1863 he presented his first public disavowal of Huxley's argument in *Man's Place in Nature* (the work in which Huxley attacked Owen) in Melbourne in the University Museum, designed on the model of the new Oxford Museum in which Huxley had debated Wilberforce three years earlier.[13] On 20 July the editorial in the usually pro-Darwinian newspaper, the Melbourne *Argus*, severely criticised the intemperance of Halford's rhetoric on that occasion. He had claimed, foolishly as it turned out, that Huxley's book was 'one which a devil might have written'.[14] The *Argus* lambasted Halford:

> What have our lecturers to do with the devil or his works? ... The devil has no business in a scientific lecture room ... Let him [Halford] keep his theatrical properties under lock and key when discussing a question of pure science ... No university of Europe would tolerate such language as Professor Halford has used of Professor Huxley.[15]

It was not only the *Argus* that was critical of Halford. *Melbourne Punch* used his outburst as material for a series of jokes:

> People I have never met: The Devil, except at Professor Halford's;
> Why is Professor Halford like a Chinaman examined in Court? Because the Professor talks of a Devil, and the Chinaman swears by Lucifer;
> New Publications: *Historical Parallels, Huxley and the Devil* by Halford;
> Mr Punch yesterday, required some stimulating diet, asked a waiter at the Café to provide him with a devil. The obliging servitor replied at once – "A Halford, sir, yes, sir." Such is fame![16]

Melbourne Punch also published an extended parody of Halford's lecture in the form of a lecture by a 'venerable gorilla' to a collection of apes, in which he shocked them by suggesting that 'infamous attempts were being made by individuals belonging to the human family to establish a relationship between it and the simious tribe'.[17]

As a concession to Huxley, Halford decided to take the debate concerning humanity's origins to the anatomy of the hands and feet of hominids and apes. As one of Halford's successors in the Chair of Physiology, William Osborne argued: 'When he found he could not support Owen in his famous controversy with Huxley over the comparative anatomy of the brain, he honestly admitted his position; but he attacked Huxley on the detail of hand and foot anatomy.'[18] The debate was 'a glass half full or half empty' kind of event. Halford claimed in public lectures and in his publications, most notably in *Not Like Man, Bimanous and Biped; nor yet Quadrumanous; but Cheiropodus* (1863),[19] that while man had two hands and two feet, the same could not be said of the apes, as there were many hand-like anatomical features in their so-called 'feet'. Huxley and other Darwinists merely side-stepped Halford's criticisms, saying that one did not require all features to be the same in hominids, merely that enough shared characteristics developed over time indicated a shared ancestry. Also, important shared characteristics were defined as 'primary' while those that supposedly did not matter were described as 'secondary', an easy way of dismissing any conflict in the evidence. As both sides argued from different premises, the debate was intellectually a damp squid—if you believed in evolution then the 'foot' was half foot-like with vestigial features of the hand. If you did not accept evolutionary theory, the 'foot' was more hand-like and, as it had been created for eternity, was not an example of the evolutionary process. It was not about to become more foot-like, nor did it illustrate evolutionary processes. The protagonists' arguments were based on totally different taxonomic approaches. Such problems in distinguishing between species were not new in such circles, nor are they resolved today. Well before Darwin, the English philosopher John Locke – a great proponent of the idea of the Great Chain of Being – wrote that he was unable to see 'why a shaggy-haired dog and a hound are not as distinct species as a spaniel and an elephant' and that 'the boundaries of species, whereby men sort them, are made by men.'[20] The debate was further devalued due to the fact that both sides compromised their scientific respectability by a distinct lack of accuracy when referring to apes or macaques. Both terms were used interchangeably without reason or explanation and thus created confusion.

Picking a fight with Huxley was a mistake. Halford was to share the same fate as Huxley's British opponents, initially through the agency of a review in the English medical journal, *The Lancet*. The introduction to the review began with the observation that:

> There is something so novel in a contribution to anatomical science from the antipodes, that we are disposed to give this first emanation from the new Professorship of Anatomy and Physiology in the rising University of Melbourne a hearty welcome.[21]

After a lengthy analysis of Halford's argument from a pro-Huxley stance, it summarised Halford's contribution contemptuously:

> We cannot conclude without expressing our regret at the publication of this pamphlet, as, independent of its scientific deficiencies, its style and tone are such as, with many readers, to invest the subjects of which it treats with an importance entirely foreign to their real nature.[22]

In a strange twist which smacks of dishonesty, and to make certain that none of Halford's anatomical criticisms of Huxley's analysis had any effect, it concluded:

> investigations into the nature of this mysterious "humanity", so difficult to define, ... belong rather to the province of the metaphysician than that of the anatomist. Those who hold it in high esteem will not be satisfied with distinctions drawn from muscles and tendons.[23]

Ironically, such an attack brings to mind Huxley's criticism of Wilberforce in 1860, when he said of Wilberforce's sarcastic rhetoric that he would prefer an ape as an ancestor rather than an intelligent scientist who used insults to attempt to bury legitimate argument. Osborne wrote of this attack on Halford: 'that there was something in his [Halford's] contention is to be shown by *The Lancet's* four columns of hostile and not too fair criticism.'[24]

Huxley supported Halford's opponents in Melbourne and they corresponded with him. As Huxley's biographer, Adrian Desmond, has

Dispute between William Thomson and Halford in *Melbourne Punch* 1865.
Courtesy: State Library of Victoria.

written, 'He had them fighting in Melbourne'.[25] Halford's most significant Australian opponent was the brilliant William Thomson, later to take up the public advocacy of germ theory with as much ferocity as he did Darwinism.[26] Thomson had been a public critic of the early Medical School and defended Huxley from Halford's attacks under the pseudonym of 'Opifer' in the *Argus*. Halford took little time to discover the identity of 'Opifer' and, when Thomson arrived at the Anatomy Theatre in the University to take up a general invitation to examine the monkey that was the main exhibit in Halford's public lecture, Halford was ready and waiting. Halford 'demanded what he (the supposed 'Opifer') had to say [as to] why he should not be knocked down'. 'Scalpel and forceps in hand' Halford drove out Thomson and his companions.[27] Such behaviour did little for Halford's argument. As a sign of his fall from grace, Huxley referred to Halford disparagingly as 'your eccentric professor of medicine at the University of Melbourne'.[28]

If this was indeed the School's first foray into the world of international research, it was a controversial beginning and a litmus test for colonial acceptance in the greater world of the British Empire. Four years later, in 1867, Halford optimistically applied to the General Medical Council to have the Melbourne degree recognised in Britain. He was unsuccessful. Whether this was due to his position on Darwinism or the unproved quality of the course is not known.

After his initial intemperate assault, Halford offered more measured criticism of evolution, although whether this change came about because of the public criticism in 1863 or a change of heart is difficult to ascertain. Later he based his criticisms on the lack of compelling concrete proof for Darwin's theory. His attitude had become acerbic rather than crusading: 'One visiting lecturer who spoke at large on evolution, drew a rejoinder from Halford ... in defence of Professor Owen: "Hyperion to a satyr. Solid science to windy nonsense."'[29] Those who remembered him were inclined to place the bulk of his criticism into the latter category. A L Kenny wrote that 'in the whirlpool of the evolutionary controversy he was not Darwinian; he was independent as was his great instructor in comparative anatomy, Professor Owen.'[30] There was no shame or error in taking this stance as Darwin failed to answer most of the trenchant contemporary criticisms of his theory, although evolution was robust and its defenders expected to solve the problems with mechanism in the fullness of time. As I noted earlier, Darwin downplayed natural selection, the biggest hurdle to widespread acceptance of evolutionism, and reverted to Lamarckism in his later years.[31]

In his lectures to medical students Halford presented six objections to Darwin's theory.[32] These include legitimate criticisms and observations that were eventually shown to be incorrect, especially in relation to the order of appearance of related species.

Melbourne's professors in retreat

Unfortunately for Halford the supporters of Huxley and Darwin, while not necessarily accepting of Darwin's ideas, expected all to bow to the new orthodoxy of Darwinism. Even Huxley never really accepted the idea of natural selection, though he did not acknowledge this publicly.[33] Huxley and his ilk were imbued with the race and class attitudes of their time and they saw direction and progress through such eyes. It was obvious to them that the English middle class was the pinnacle of evolution and must have reached its pre-eminence through innate superiority. Such an idea had deep roots and survived until the Second World War at least. As Dr S V Sewell (later knighted) put it to the Australian Branch of the British Medical Association Congress in 1923, some sixty years after Halford's debate with Huxley: 'It is a big step from the gesture language of the lowest type of savage to the extensive word and phrase vocabulary of that finest product of evolution, the educated upper middle-class English gentleman.'[34] The consequence of the victory of Darwinism was not that it was necessary to agree with all of Darwin's theories, just that it was important not to air criticisms in public. Those who were out of line were vilified and excluded from preferment. Owen and St George Jackson Mivart were among those who suffered in England, although Owen did get the position he coveted at the Natural History Museum.[35] Mivart, a Roman Catholic, had been an evolutionist when studying with Huxley, but upon obtaining a Chair with Huxley's help, began to express his doubts about natural selection and the view that purely natural processes could explain the development of life. He argued that no matter how simple the earliest eye was, it must have had purpose and direction in its development in the first instance. The fourth Professor of Anatomy in Melbourne, Frederic Wood Jones, would resurrect such ideas in the twentieth century.

Halford's defence proved to be a rearguard action. The scope of the eventual defeat of McCoy and Halford was such that, the Melbourne *Age*, initially violently opposed to Darwinism, editorialised the day after Darwin's funeral in 1882 that 'modern science is … completely penetrated by the Darwinian view'.[36] Furthermore, although perhaps uncommitted in the evolution–creation debate, wider middle-class opinion in a wide range of public policy matters came to be influenced by a social Darwinist world view.[37] As the *Pall Mall Gazette* observed on the day of Darwin's funeral, 21 April 1882:

> The Darwinian creed ... tinges our unformed public notions; it reappears under a hundred disguises in works of law and history, in political speeches and religious discourses, in artistic theories and vague social speculations ... If we try to think ourselves away from it we must think ourselves entirely away from our age.[38]

Social Darwinism in all its manifestations also became increasingly popular among Australian intellectual circles. In 1874 the future Prime Minister of Australia (sometimes known as the father of the Liberal Party), Alfred Deakin, lectured a group of fellow students in the quadrangle of the University of Melbourne Law School, explaining that the social Darwinism of Herbert Spencer had superseded all traditional subjects such as Classics and Mathematics. He told them that the 'world was entering on a new epoch in history: everyone should start anew from Spencer's theories'.[39] Social Darwinists saw the State as an organic entity that followed Darwin's laws of evolution. The most radical consequence of this theory was that the State should play only the most minimal role in social and political life. Thus Spencer advocated the abolition of the Poor Laws, state banking and national education as well as the established church. He also wanted the lifting of any restrictions on commerce and the abolition of factory and public sanitation legislation. In Spencer's view, the weaker members ought to go to the wall for this was the only way that society could improve. Spencer coined the term the 'survival of the fittest'.

Halford and protoplasm

Another reason for doubting Halford's wholesale support of a simple creationism was his attitude towards protoplasm. In February 1869 Huxley gave a lecture (later published and widely circulated) titled 'On the Physical Basis of Life',[40] which proved to be an important moment in the great battle between vitalism and materialism—a battle that would be taken up by Frederic Wood Jones in Melbourne in the following century. In his lecture Huxley introduced the idea of protoplasm, his solution to the creation of life, being the material of which all living things were composed. No longer needing a 'spirit' to animate thought and soul, humanity, so he argued, was made of the same sort of primeval slime as bacteria. In response to Huxley and about five months later, Frederick McCoy lectured to a packed auditorium of over seven hundred spellbound Melburnians. The *Age* reported that in the course of the lecture he argued against:

> some of the recent theories as to the basis of life and the succession of animated beings. The protoplasmic hypothesis of Huxley to account for the former, and the natural selection of Darwin for the latter were both dealt with trenchantly ... Professor McCoy ... pointed to his own collected specimens as a standing refutation of Darwin.[41]

A few days earlier McCoy had told the *Argus* readers that a viewing of the gorillas in the Museum was the perfect antidote to counteract the Darwinian poison:

> I am perfectly sure no student of zoology and comparative anatomy at our university, nor student in our medical school, can ever hold such views, now that immediately after lectures the characters can be demonstrated from nature.[42]

Those gorillas were at it again. Halford had also asked to speak on protoplasm but was refused permission by the University Council. At first it would seem that it was his ill-starred public performances in the previous six years that worked against him. Bishop Perry had urged the Council to ban Halford's lecture.[43] It is striking that it was the cleric on the University Council that opposed Halford, especially considering Halford's defence of creationism. He must have known that Halford's views were not as conventional as they had been in 1863. Halford did, however, find another outlet to publicly air his views on protoplasm. This occurred on Wednesday 30 June 1869 when the Reverend Mr Higginson gave a lecture attacking the idea of protoplasm to an audience of two hundred at the Mechanics Institute. The *Leader* report claimed that Higginson's attack on protoplasm was mediocre, 'advancing no arguments in refutation' and that in conclusion, the 'lecturer resumed his seat amidst some feeble attempts at applause'.[44] This uneventful meeting was dispersing when Halford jumped to his feet, declaiming that the views that had been attributed to him by the lecturer were false and:

> though he had differed widely from Professor Huxley in the matter of "Man's Place in Nature", yet he must support him in the present instance. He (Professor Huxley) had taken the only view which a scientific man could take.[45]

At this, the meeting descended into an 'uproar of mingled applause and hisses, during which Professor Halford ceased to address the meeting'.[46] It was then reported that an impromptu meeting of Huxley's supporters made arrangements for Halford to give a public lecture, but this did not eventuate.

Darwin's triumph – a white Antipodean race

While Halford and McCoy held out for creationist theory in their lectures and public statements, Darwinism eventually spread throughout the University as new appointments were made, beginning with Walter Baldwin Spencer as Professor of Biology in 1887.[47]

There was another important reason for the passion aroused in this debate. All agreed that populating and moulding the new continent (or *tabula rasa*) was a priority. It was this belief that led to the creation of the Zoological and Acclimatisation Society of Victoria, established in 1862. Its aim, as articulated by the first president, McCoy, was to bring civilisation to a savage land. This project was endorsed by numerous scientists, such as the botanist, Ferdinand von Mueller, and the statistician, William Archer, as well as many interested worthies, such as Edward Wilson, editor of the *Argus*, a contributor to Darwin's research and later in life, a neighbour and friend in Kent. As we shall see, Wilson's Trust, founded after his death, funded Richard Berry's anthropometric laboratory at the Children's Hospital in the 1920s, set up for the purpose of coordinating the breeding of a better race. To accomplish the transformation of Australia, the Zoological and Acclimatisation Society of Victoria saw as its task 'the perfection, propagation, and hybridisation of races newly introduced or already domesticated'.[48] From its first base in the Botanic Gardens in South Yarra, and later the Zoological Gardens in Parkville, numerous imported animals from similar latitudes around the world were taken and released into the wild. The number and variety of different species released into the Australian hinterland is quite remarkable.

The economic consequences are still being felt today, including attempts to solve the periodic rabbit plagues that have degraded the Australian landscape. Rabbits were released by the Society in Victoria in 1863. The Society's belief in science as a tool to manipulate and create a better environment for Europeans also extended to the planned destruction of native species. The Honourable Thomas Turner a'Beckett, member of the Legislative Council, proposed to the first meeting of the Society, that because 'the English song-birds which had been set free in some parts of the Colony were much persecuted by the native hawks ... [the hawks'] extirpation would therefore be an advantage'.[49] Perhaps the Australian environment we see today in the temperate zones is no more than a giant nineteenth century Darwinian experiment gone wrong—a Dr Moreau's Island or a Jurassic Park in the Antipodes. The scientific interests of the Zoological and Acclimatisation Society of Victoria went further than the release of strange animals and plants. The obverse side of acclimatisation was the concern about whether English bodies could successfully survive unaltered in Melbourne's climate.[50] As early as 1873 the Society was recommending a southern European way of life for the people of Victoria.

As far as populating this new land was concerned, many writers eagerly awaited the product of the racial experiment. Doctors had an important role to play. As Anderson has written, 'the climate was foreign, social life appeared disordered, the diseases varied, and it sometimes seemed that a new biological type might emerge from the colonial turmoil.'[51] Marcus Clarke in *The Future of the Australian Race* (1877), ruminated on how the Australian environment would shape the Australian 'racial type'. Fear of 'degenerating' to an Aboriginal type was a paramount concern. The *Herald* expressed this popular view in an article in September 1934 when discussing a prospective lecture by Dr John Cumpston (the Commonwealth Director General of Health and a graduate of the Melbourne School) at the Institute of Anatomy in Canberra on the development of the Australian people as a chapter in the medical history of mankind:

> From time to time discussion has risen as to the physical appearance of the typical Australian after the lapse of a few centuries. Will climate and other influences lead to the development of a race differing greatly from the stock of the British Isles? Will the Australian be "tall and freckled and sandy", the true Cornstalk, or retain the physique of his forefathers? ... the Federal

> Director General of Health ... should have something interesting to say on this subject next week ... I hope the doctor does not lean to the opinion ... that Australians would gradually assume the colour and other physical attributes of the aborigines.[52]

Many medical students reflected in a Darwinian manner on this question. In 1891, 'studentia' wrote in the Medical Students' Society journal, *Speculum*, that 'Anyone who visits our National Museum cannot fail to be struck with the resemblance between the specimens of the higher apes and our aboriginal brethren', although he believed that 'Poor Jacky Jacky would turn away from his hairy ancestor with about the same feelings of contempt that one of our patrician ladies would show, if her aboriginal brother claimed relationship'. He concluded that 'Taking all into consideration, there is nothing so very repulsive after all when we look back to our progenitors. It is comforting to feel that we are *evolved* from them'.[53] In 1894 in 'A New View of Evolution' (again in *Speculum*), 'W. L. F. M.' told the readers that 'I spend most Sundays at the zoo in contemplation of the vast evolutionary process which has culminated in the race of which the present writer is a member. To put it more concisely, I spend that day in visiting my poor relations.'[54]

In the Medical School, the first Australian-born professor at the University, Harry Brookes Allen, appointed in 1882 as Professor of Descriptive Anatomy and Pathology, eventually became the Dean. Allen's reign instituted a less controversial period in the history of the teaching of anatomy; in the public domain at least. After Halford and McCoy had both passed away, Allen acknowledged in 1914, the debt owed by science to Darwin, at the celebration of the first fifty years of the Medical School. Though he never attended church, like almost all Victorian gentlemen, he found it difficult to cast God aside, and accept in His place the notion of evolution expressed through random, natural selection. He wrote in his poem 'Evolution and the Ideal', published in 1899:

> Thus life appears a vast ascending scale,
> Wherein the strong must constantly prevail;
> And every creature in its growth displays
> In brief the story of the earlier days.
> ...

> Hence dull equality is deadly foe
> To upward growth, while variations show
> The play of weak and strong, with promise sure
> That evermore the fittest shall endure.
>
> While all the air with harmony resounds
> From tribes and tongues to Earth's remotest bounds,
> Praising the God who gave mankind to rise
> Out of the realms where lower nature lies.[55]

The Great War of 1914–18 was to challenge that difficult, though easily worn, compromise. Many then and later saw the ideas of an unremitting and heartless social Darwinism and its allied biological racism as a major cause for the human catastrophe of that event.[56] This caused some to reject the materialism of Darwinism. As we shall see later, the fourth Professor of Anatomy, Frederic Wood Jones, was such an individual. He wrote that it is possible to find:

> In the teaching of Huxley ... those seeds of thought which, broadcast among the uncritical masses, developed into the malign mentality that dictated the tragedy of 1914.[57]

CHAPTER THREE

Anatomy and the rise of the Medical School, 1880–1905

On the evening of Tuesday 7 April 1891, the Melbourne University students held their first 'Students' Night' of the year. As it was a fine evening, the students, resplendent in academic dress, decided to march to the Royal Exhibition Building on Nicholson Street, Carlton Gardens, to take part in the 'People's Palace'. This was a huge bazaar, organised by various church congregations in the South Yarra district.[1] The Exhibition Building was a perfect venue, being the largest covered space in the city. It had been built so that Melbourne could take part in that great outpouring of national pride and progress among Western nations, the International Exhibition Movement. Only three years previously the second exhibition had been an enormous success, showcasing Melbourne's reputation as one of the leading cities in the world. Unlike the uncertainty and pessimism of the 1860s and early 1870s, such confidence was also reflected in the Medical School. Two years later, the disastrous bank crash that precipitated the decade long depression in Victoria dented that confidence, severely inhibiting the growth and ongoing revival of the Medical School. Before the crash, however, in 1884, the editorial of the Medical Students' Society journal *Speculum* on praising the Medical School, asserted:

> In writing on this subject we are likely to fall under the condemnation levelled against Australians by the late Mr [Anthony] Trollope – that of an aptitude for "blowing" [boasting]. However, as we are proud of the rise of our school and the ability and energy of its professors [we believe that]... Never in the history of any Medical School in the world, so far as we know, has the rise and progress been so signal and so sure.[2]

From its uncertain and slow start, the Medical School in the 1880s and early 1890s was by far and away the dominant faculty in the University, numerically and financially. It had also developed a strong sense of individuality, symbolised by its location across the lake from the other faculties housed in the University's original cloistered building. This separation was originally adopted because of the University Council's perception that the dissecting room would offend 'non-medicals'. By the 1880s, therefore, medical students referred to the rest of the University as the 'Other Side'— the domain of the 'Intellectuals'.[3] Antagonism between the two groups facing each other across the lake festered for decades, occasionally exploding into outright hostility. As 'G. J. N.' wrote in *Speculum* in May 1887, 'the medical students as a body have not ... shown any inclination to extend their sympathy and support to the "other side".'[4] In 1886 the Medical Students' Society (founded in 1880) had flexed its muscle and taken on the University Council – to the extent of personally petitioning, and meeting with the Premier of Victoria – to obtain a greater share of University income on the basis that they made up half the number of students, and yet received less than one-ninth of the government grant.[5] The victory of the Medical School in this matter was greatly assisted by student action.[6] Also, medical students, though frequently complaining of their heavy workload, considered themselves to be more convivial than the 'Intellectuals'.[7] Hence the procession to the Exhibition Building was dominated by medical students, as was the performance upon their arrival. It was claimed that popular opinion had it that 'next to a worker in a brewery, a medical student stands first as a complete absorbent of fermented liquors.'[8]

At 7.30pm on the evening of the procession to the People's Palace, students from all faculties and colleges assembled behind the medical students at the Medical School in the following order: University Union, Law, Engineering, Ormond College, Trinity College and Queen's College.

They proceeded to march along Elgin, Lygon, Victoria and Nicholson Streets to the Exhibition Building:

> It was a charming night, the clerk of the weather being for once generous. About four hundred students put in an appearance, at 7.45 a start was made, the brass band, under the leadership of our old friend, Tom Linehan, striking up the University Anthem. *En route* various selections were played by the band, and the students sung their now well-known "varsity songs." Police protection was kindly provided ... so there was no interference, other than some good-natured banter between town and gown, to in any way mar the procession.[9]

As we shall see, this was not always the case, as antagonism existed not only between the two university groups, but also between 'medicals' and 'town', particularly the working class residents of the slums of Carlton, Fitzroy and North Melbourne, suburbs adjacent to the University.[10] In fact during a torchlight procession in 1900 (also from the University to the Exhibition Building and organised to celebrate the relief of Mafeking),[11] the medical students were assaulted by gangs of larrikins 'who quickly found out that the students were not to be trifled with'.[12] It must have been a colourful battle as the medical students that night were in fancy dress, including Cupid, a hospital nurse, John Bull, a bubonic rat, a plague inoculator, Venus, Death, Baden Powell, a Red Indian, a Geisha, a corpse and Yankee Doodle.

After the earlier procession to the People's Palace in 1891 and upon arrival at the Exhibition Building, the students set themselves up on a temporary dais and performed popular songs for over an hour with the conductor wielding a femur (thigh bone) rather than a baton. Throughout the concert, the students' mascot – a skeleton dressed in cap and gown with a false pink nose – performed antics for the entertainment of the audience. It was during the singing of *Old John Brown*: 'that the real time of the skeleton came in, as it took entire possession of the platform and during the "dumb" verse *Brown's sole* truly went marching on, to the intense amusement and delight of the large audience assembled'.[13] After the singing had concluded, many of the audience attired in fancy dress joined the robed students in a 'March of the Nations': 'the unique costumes of the ladies and gentlemen, intermingled with the academical

[sic] robes of the students, and various parts of different skeletons, made a very pretty spectacle.'[14]

The femur and the skeleton as symbols of the medical students became ubiquitous. The Medical Students' Society had adopted the crossed femurs and skull – the 'Jolly Roger' – as its emblem, which it displayed at its meetings in the form of a statue and, in public appearances, on its banner.[15] One of the first and most memorable public demonstrations of the new symbol was the mass attendance of the medicals at a performance of Shakespeare's *Macbeth* on Saturday 9 August 1884, starring Miss Genevieve Ward at the Theatre Royal.[16] The Medical Students' Society journal reported that after assembling in the gods (the highest most seats) and announcing their arrival in song: 'cheers were redoubled when the piratical emblem "our own flag" consisting of a forked banner with the device of a huge skull above two crossed femora (thigh bones) *blanc* upon a field *noir* was let down.'[17] The use of the skeleton as a symbol of medical study was telling. In a period when university trained medical practitioners were competing with a large array of alternative medical operators, one way in which the University claimed orthodoxy and superiority was in its detailed and exhaustive study of anatomy. This was the point of superiority continually raised by the university trained medicals. As 'Rotomahana' versified in 1897 in *Speculum*:

> And in the Second Year you slake your thirst.
> Here 'tis Anatomy wins your affections,
> With evil-smelling bones and worse dissections;
> ...
>
> And then the Third comes in and soon is sped.
> Anatomy still holds her rigid sway,
> With subjects in all stages of decay.[18]

In 1897 the Medical Students' Society, discussing the study of 'all-important Anatomy', asked 'if our men are to retain the high name that they have heretofore borne as skilled anatomists',[19] there needed to be a much more regular supply of cadavers for dissection.

It should be said that University trained doctors in the nineteenth century had few, if any, therapies for the sick that were consistently and demonstratively better than those of unorthodox practitioners. The pharmacopoeia taught at the University contained none of the magic bullets of late twentieth century medical practice. The first such drug that was aimed specifically at a disease-causing micro-organism was salvarsan, a relatively ineffective treatment for syphilis, discovered by Paul Ehrlich in 1910. Phlebotomy, or bloodletting (one of the standard practices of the old humoral medicine), persisted as a therapy until the close of the nineteenth century, even though Pierre Louis had shown in Paris early in the nineteenth century in *la méthode numérique* that it was a useless treatment.[20] Although many of the old drugs continued to be used in practice, there was a good deal of doubt concerning the efficacy of many of the most commonly used therapies. In an article in the *Australasian Medical Gazette* of 1 January 1901, the old controversialist and Melbourne University Lecturer in Forensics, James E Neild, listed those practices that he believed had finally been discarded by the profession:

> Thus, we do not carry about us a case of lancets to phlebotomise [bleed] every other patient whom we visit; we do not incessantly cover the temples, the thorax, and the abdomen with those hideous black blood-suckers—to wit, leeches; nor do we everlastingly clothe our patients with cantharides plasters; nor do we, on the smallest provocation scarify the skins of unoffending persons and clap upon the bleeding wounds a glass fiend to drink the blood there from flowing; nor do we exhibit tatar-emetic to patients half dead with pneumonia; nor do we subscribe black draft in half-pint doses, nor drachms of calomel, nor half ounces of rhubarb, nor drachms of aloes, nor pints of infusion of gentian, nor *ad libitum* supplies of various tinctures.[21]

Neild, as much as anyone, represented the spirit of the Melbourne Medical School. Apart from taking part in Halford's initial dissecting class at the University, he had lectured in forensics almost continuously until his retirement at the age of eighty in 1905.[22] In this address Neild was mirroring the physician and Harvard professor, Oliver Wendall Holmes, who had remarked a few years earlier:

> If the whole *materia medica* [drugs], as now used, could be sunk to the bottom of the sea, it would be all the better for mankind ... and all the worse for the fishes.[23]

Neild did not mention that no real alternative therapies had been discovered to replace the old ones.[24] In 1892 he had given a lecture to the students on the contemporary view that it was accepted clinical practice that nature should be given more scope to affect healing. This was the increasingly popular response to the discrediting of most therapies; especially the heroic method of giving massive doses of medication to produce violent, noticeable physiological reactions—kill or cure. He also 'properly extolled the attention now given to the science of hygiene.'[25] The old method of inoculation for smallpox, and later rabies and then diphtheria in 1890, continued to be the only effective treatment, and it seems that the success of the diphtheria antitoxin has been exaggerated.[26] Robert Koch's much vaunted tuberculosis 'cure', tuberculin, had failed spectacularly because it increased rather than decreased the virulence of the attack. John William Springthorpe, the lecturer in therapeutics at the University, famously conducted the first clinical trial at the Melbourne Hospital on 19 March 1891. This was most likely following advice from Harry Brookes Allen, who had been present in Koch's laboratory in Germany when the breakthrough was achieved.[27] By September 1891 the students could joke that:

> By order of "Springy", Koch's gravy
> On the menu no longer appears;
> But still he conspires with De Bavay
> To discover *microbes* in our beers.[28]

As well as being the bacteriologist at the University, August de Bavay was the brewer at Fosters Brewery in Melbourne from 1894 to 1904, and was credited with inventing the Australian style lager or amber nectar.

The ineffectiveness of much of the *materia medica* continued until after the Second World War. Roy Porter quotes the distinguished American physician, Lewis Thomas, speaking of his student days in the 1930s:

> We were provided with a thin, pocket-size book called *Useful Drugs*, one hundred pages or so, and we carried this around in our white coats when we entered the teaching wards and clinics in the third year, but I cannot recall

Harry Brookes Allen, Professor of Descriptive and Surgical Anatomy and Pathology, 1882–1905. Courtesy: Medical History Museum, The University of Melbourne.

> any of our instructors ever referring to this volume. Nor do I remember much talk about treating disease at any time in the four years of medical school except by the surgeons.[29]

Contrary to the practice of medicine conducted by physicians, by the late nineteenth century, surgery, once the poor cousin in the medical family, could claim considerable progress through the agency of asepsis (cleanliness) and anaesthesia. Listerism (the use of carbolic spray to disinfect) had been first demonstrated in the Melbourne Hospital in 1882.[30] As elsewhere, the discovery of the benefits of anaesthesia encouraged surgeons to perform previously unimagined, major internal operations. Without adequate knowledge of infection and germ theory the results were catastrophic. As James Barrett (sometime demonstrator in Anatomy and later in life Chancellor of the University), pointed out in his Halford Oration in 1932, antisepsis took some time to be accepted in Melbourne:

> in 1882, whilst Sir George Syme and I were house surgeons at the Melbourne Hospital, a dramatic event took place. The hospital was supposed to be a plague house, because so many injured patients became infected with sepsis, often fatally ... while we were in residence no patient whose abdomen was opened recovered. The most elaborate precautions were taken in the way of clean rooms, fresh paint *etcetera*, but the things that really mattered escaped notice, namely the hands, clothing and instruments of the operator, because no one at that time knew what surgical sepsis was, and no one guessed that infection was caused by these agencies... at this junction a young surgeon, the late Bernard Newmarch, who had been a house surgeon to Lister, passed through Melbourne and showed us how to disinfect wounds.[31]

The study of anatomy was central to the practice of surgery, and the demonstrators in Anatomy at the University of Melbourne in the latter decades of the nineteenth century were among Melbourne's leading progressive surgeons,[32] thus forming a close collaboration that was to continue indefinitely. Therefore it is not surprising that Anatomy had such a high status, as not only was it central to the practice of surgery, it also conferred the status of the 'expert' on university trained general practitioners.

Medical students 1879. Note the skull and cross bones symbol of the Medical Students' Society.
Courtesy: Medical History Museum, The University of Melbourne.

The continued development of the Medical School was further challenged by the state of the general population's health. In Victoria, the good news concerning the health of the people finished at the operating theatre doors, as statistics dealing with the morbidity and mortality of the population seemed to inspire pessimism rather than hope. In England, and by the end of the nineteenth century, there had been considerable improvement in the health and life expectancy of the population, albeit starting from a low base. It is now clear that this was almost exclusively due to improvements in living standards along with public health measures, such as improving sanitation and housing. This revolution,

In fact this problem was the subject of the inaugural 1884 edition of *Speculum*, reporting on a long discussion that appeared in the *Argus* comparing Colonial and British degrees, in which it became clear 'that the [Colonial] students are quite as well taught in the essentials of a medical education, but have not the opportunity of specializing to the same extent'.[49] In 1885 it was proclaimed that: 'Judging from the experience of the last few months alone, one would not be led to consider that a brilliant future lay before our school or its members. The clinical teaching has never been in so unsatisfactory a state as it is at present.'[50] This was just one of a plethora of complaints concerning clinical education in the first half century of medical education at Melbourne.

As well as the lack of clinical education, the quality of teaching in hospitals was a constant concern. In 1886 a student asked:

> why is the attendance of certain members of the honorary staff so erratic? It seems so funny to hear a student at the end of his years' work say "I wish one of you fellows would point out my honorary to me, I want him to sign my certificate." It's just a little too bad to be compelled to pay fees to these gentlemen for their instruction.[51]

It is not possible to deal with this topic at any length, other than to mention that initially clinical teaching was done in an almost unsupervised manner, outside the control of the University. The students also had to pay for clinical training, which was an added expense on top of their university fees. Each student had to obtain a certificate from the hospital in the requisite clinical subjects to be eligible for a pass. As mentioned above, clinical teaching often occurred before the requisite theoretical subjects had been taken, making it, in the view of many students, irrelevant.[52] In fact, it was the Medical Students' Society that organised many of the clinical classes, especially at the Children's and Eye and Ear Hospitals.[53] The chance of getting any good hands on training was hit or miss. The students complained bitterly until, and after much tinkering, it was partially solved in 1900. Further reforms occurred in 1906 and 1908, when the University finally took full control of clinical teaching.[54]

There were also complaints in the first two decades of the course because third year Anatomy merely required a repeat attendance at second year Anatomy classes. As a 'third year man' complained in 1885: 'But the anatomy bell rings. I must do my second year work again.'[55] After this the complaints were heeded and a separate course for third year students was instituted.[56]

It was not the Medical School that was blamed for most of the trouble, but rather the University Council, which became the butt of extraordinarily strong criticism. Initially the complaints of the medical students related to funding. Although they provided half the fees and attracted half the government grants, they received far less than that expenditure. In 1886, for example, it was one-ninth of the government grant.[57] The reason *Speculum* gave for this inequity was that the Council (who allocated the money) was elected from the Senate, made up of all those who had obtained a Master of Arts. In all other courses, a single third class honours in the final year of a three-year BA allowed you to pay a sum of money and upgrade to the MA. Medical students were virtually excluded from the Senate, even though they had completed five years of study. To qualify they needed a Master of Surgery or a Doctorate in Medicine.[58] By 1884 no local medical graduate had qualified for the Senate, thus being ineligible for election to the Council. The medical membership of the Senate consisted solely of graduates from abroad who had been granted honorary doctorates. And they made up less than ten per cent of the Senate.[59] To make matters worse, the Medical Students' Society, of which Professor Allen was ex-officio chairman for most of its first twenty years, was highly critical of those medical men who were on the Council, claiming that they had no knowledge of or interest in the Medical School. As *Speculum* noted in 1886: 'as of the three medical men in the body, only one takes the slightest interest in the place, the other two not having been near our school for years, and having the most complete ignorance of the curriculum.'[60] This is a somewhat enigmatic statement because there were, in fact, five medical councillors. Apart from Brownless, who was not elected by the Senate and was strongly supportive of the Medical School (having played the major role in its foundation), the other four medical men on the Council in the latter part of the nineteenth century were the mesmerist and spiritualist, James Motherwell; the astronomer and former surgeon, Robert L J Ellery, FRS, FRAS; William Henry Cutts MD, appointed in 1858; and James Robertson MA, MD, appointed in 1885 and a lecturer in the Medical School. It seems Robertson was the only one favoured by the medical students, though

Motherwell possessed the sort of progressive streak they would have favoured. Ellery was openly criticised by the students and Cutts was also criticised by implication.[61]

Despite all the complaints, and as I have previously noted, the one aspect of the curriculum that was always held up as distinctly 'Melbourne' and as evidence of the high quality of the course was the extensive study of Anatomy. But how good was the teaching of Anatomy, and what were the major problems encountered by anatomists and students? There were three substantial problems: the severe difficulties experienced in obtaining sufficient cadavers (which deserves a chapter of its own), the quality of teaching aids, such as anatomical models and specimens, and the quality of teaching and supervision.

The teaching of Anatomy, 1863–1905

Upon his arrival in Melbourne on an 'infernally hot' summer's day in February 1906, the new Professor of Anatomy, Richard Berry, was met on the boat by two Melbourne anatomists, George Rennie and Colin McKenzie, who were both teaching at the Medical School and would be colleagues of Berry's in the Anatomy Department. They expressed their hope that Berry would get 'Anatomy ... out of the bog in which ... it had too long wallowed in Melbourne.'[62] Not one to waste time, Berry immediately took himself to the University to ascertain whether the grim tales of the Anatomy Department were true. According to Berry:

> it was worse. It contained literally nothing, not even a skeleton ... [t]here were no diagrams, models, osteological specimens, dissected parts for study, lantern slides, microscopes, no books, no anything, not even a room for the professor to sit in.[63]

The dissecting room also aroused no greater admiration. Berry later wrote that it was 'somewhat unusual. A long bare room with some boards on trestles. These I was told were the dissecting tables. Nothing marble topped about them. Nor were the massive awkward seats around the tables altogether what I had been accustomed to'.[64] Also the room bore the mark of 'meat fights' that enlivened dissecting classes.[65] Had the teaching of Anatomy been wallowing in a 'bog' for a lengthy period, and if so, why? On reading Berry's description of the state of the discipline of Anatomy at the University on his arrival, this would have led the reader to believe he brought not only his fiery temper, but also a rare competency in that discipline. As we have already seen, anatomy was a subject of great interest in Melbourne in the decades following the discovery of gold. Both McCoy and Halford had received extensive training under excellent anatomists, but neither maintained it as their primary interest. Harry Brookes Allen, the locally born and educated second Professor of Anatomy, was primarily interested in pathology and he had been taught Anatomy, as it were, second hand, for neither of his teachers were anatomists by choice.

In the earliest days of the School, one-third of the Anatomy lectures constituted a general, watered-down course, consisting of General Anatomy, Physiology and Pathology. In Halford's course on Anatomy delivered in 1873, Allen learnt maternal impressions during pregnancy were important in cases of epilepsy and more:

> A woman with child, 3 months, was much terrified by a monkey which was attached to a street organ jumping on her back as she went by ... when her child 12 years old was admitted to the hospital the left upper extremity and the greater part of the corresponding side of the trunk of the neck were deeply stained with dark brown pigment from which grew an abundant crop of brown harsh lank hair ... the arm was long, thin and withered ... in fact the upper limb, shoulder and back bore a very strong resemblance to those parts of a monkey.[66]

It is drawing a long bow, however, to claim that no adequate teaching of anatomy had occurred in the first fifty years of the Medical School. Some of it must have been competent. When the initial dissection class took place in Australia in 1863 in a shed erected in Halford's backyard in Madelaine Street, it comprised three second year students. They were Patrick Moloney (later physician at the Melbourne Hospital, and a poet and writer of considerable talent), William Carey Rees (a future demonstrator in the Department) and Alexander Mackie (who abandoned the course to become a Presbyterian minister, then returned to medical studies later in life, but died before completion of his degree) along with two practitioners, Drs Neild and Fetherston.[67] Fifty years on the situation was quite different. In 1905, the year before Berry arrived, there were at least one hundred and ninety-five students dissecting.

Inadequate staffing and one suspects supervision in the dissecting room, plagued the early years under Halford. In 1870 William John Smith was appointed Demonstrator of Anatomy and Curator of the Pathological Museum at the Melbourne Hospital. A year later he resigned and Dr O V Lawrence succeeded him in late 1871.[68] Lawrence lasted for four years and had a poor reputation as a surgeon at the Melbourne Hospital—chronic absenteeism being his major sin.[69] It is therefore hard to imagine he was more diligent supervising dissections at the University. In May 1876, Harry Brookes Allen, although only just admitted to the degree of MB, was given charge of the dissecting rooms in the mornings. Allen quickly made himself very busy. Later that year he became Demonstrator of Anatomy and Sub-Conservator of the University Museum of Anatomy and Pathology, and the Committee of the Melbourne Hospital appointed him resident Pathologist.[70] Allen's appointment seemed to provide some stability until in June 1882, John David Thomas MD, was appointed full-time Assistant Demonstrator in Anatomy to ease Allen's workload.[71] Thomas, however, died of phthisis (tuberculosis) shortly after he was appointed and was replaced by Mr A V Henderson.[72]

Allen was appointed Professor of Descriptive and Surgical Anatomy and Pathology in 1882, and in 1884 the system changed again, with two half-time Demonstrators of Anatomy, rather than one full-time Demonstrator being appointed. They were William (Jerry) Moore (later a dour and important figure in Melbourne medical circles) and Fred D Bird (son of a pioneer doctor and academic, important surgeon and something of a lady's man).[73] Both Bird and Moore became highly regarded surgeons and, according to their popularity with the students, seem to have provided good teaching.[74] In 1887 George A Syme, MB, FRCS, succeeded Bird as Demonstrator of Anatomy. Syme was one of the leading surgeons of his era and connected to the owners of the *Age* newspaper. He was instrumental in the foundation of the College of Surgeons of Australasia in 1927. His role as a surgeon is beyond the scope of this history, but he gave eighteen years valuable service to the Anatomy Department as Demonstrator and later, Lecturer.[75] At the beginning of 1888 Moore resigned his office as Demonstrator and was succeeded in turn by C H Mollison, MB, MRCS (later pathologist at the Melbourne Hospital for

The Histology Laboratory (early twentieth century).
Courtesy: Medical History Museum, The University of Melbourne.

sixty-four years); C J Shields, MB, FRCS in 1889 (outpatient surgeon at the Melbourne Hospital); F H Langlands, MB, FRCS in 1894 (a surgeon at the Melbourne Hospital and part of Syme's brilliant coterie); and George C Rennie, MB, MS, FRCS in 1898 (yet another distinguished Melbourne Hospital surgeon).[76] Until the arrival of Richard Berry in 1906, these men carried the teaching load in the dissection room. It needs to be born in mind that all were very busy surgeons, which allowed little time for their dissecting duties. By the 1880s almost one hundred students were dissecting for long periods. It was therefore necessary to deputise the bulk of the supervision. The stories of chronic misbehaviour in the dissecting rooms, to be discussed in the next chapter, would suggest that senior students (prosectors) supervised the room for lengthy periods. Throughout this time, the constant figure was Harry Brookes Allen, who had started as a Demonstrator in the Anatomy course in 1876, and became Professor of Anatomy and Pathology in 1882.

Unlike the description given to Berry in 1906, *Alma Mater* published a glowing description of the Anatomy curriculum in 1898. The journal recorded that the initial taste of dissection for the students occurred in second year Anatomy. This course comprised Professor Allen lecturing 'thrice a week, dealing mainly with the organs, the lectures being illustrated with specimens, models and abundance of coloured drawings on the blackboard, which are faithfully reproduced in the students' notebooks'.[77] Dissections associated with the lectures were conducted under the direction of demonstrators (usually surgeons). In third year, each student performed a 'second course of dissections, working out every part in detail for himself.'[78] The remainder of the course was 'devoted to an exposition of the chief difficulties which are experienced, and the common mistakes made by young anatomists.'[79] Also, twice a week Allen conducted a class for third year students in the Pathological Theatre of the Melbourne Hospital. As *Alma Mater* reported, 'the organs are thus studied in recent state, and the vessels and nerves of the trunk and the anatomy of many important surgical regions are demonstrated to great advantage. An introduction to the elements of Pathology is gained at the same time.'[80] Another year of Anatomy was offered in the third year.

Melbourne medical students also received Anatomy classes in the fourth year. According to *Alma Mater* they were: 'confined to Regional and Applied Anatomy, with a third course of dissections, but without lectures. The examination is partly written and partly practical by actual dissections.'[81] The purpose of such an extended course was because: 'the prolonged study of Anatomy … should continue throughout the course till the student can visualise every part of the body.'[82]

How accurate is this description? To some degree this article, written in September 1898, marks a watershed in the teaching of Anatomy, because May 1898 marks the final public complaint (for a considerable period) about a lack of cadavers that had bedevilled the teaching of Anatomy and had on numerous occasions threatened to close the Medical School.[83] This beneficial change could well have been the result of the introduction of the use of formaldehyde to preserve the corpses after its discovery in 1893. From the outset, dissection had had its own term dates from May to October, so as to take advantage of the cooler weather and delay the decomposition of the cadavers being dissected.[84] Students were expected to be available to dissect, depending upon the availability of cadavers. After the introduction of formaldehyde, dissection took up the mornings of every term day for three years.[85] By 1905, *Speculum* could report that 'the old complaint that demonstrations were lacking in the Dissecting-Room can no longer be made ... the teaching afforded in this branch is improving year by year, thus affording to some degree an element none too conspicuous in the course—viz., practice.'[86] I discuss the serious lack of dissecting material in the next chapter, but Berry certainly arrived just at the moment one of the major obstacles to good teaching had been removed.

Allen: the teaching and supervision

It seems it was public knowledge that Allen did not want to teach Anatomy. In 1905 the editorial in *Speculum* announced that: 'We still look forward to a time when Professor Allen, freed from the burden of Senior Anatomy, will be able to concentrate all his attention on Pathology.'[87] Was Allen burdened with too much teaching responsibility? That was certainly the opinion of the Royal Commission on the University, which recommended the splitting of Allen's Chair and the appointment of a new Professor of Anatomy.[88]

Just prior to Richard Berry's arrival in 1906, the students published a poem concerning the Chair of Pathology and Anatomy being split, and the imagined response of Harry Brookes Allen:

> Fair maid Pathology,
> They wrong thee much,
> To rot resembling thee,
> Or double Dutch.
> Thy beauties I extol,
> Thy spirit meet;
> With methyl-alcohol
> Thy breath is sweet.
> And I have cherished thee
> Through all my life;
> Yet was Anatomy my wedded wife.
> My vows another claimed
> With lawful bond;
> Of thee I was ashamed

To seem too fond.
But now I'm free to vent
My love's full force;
For, thanks to Tommy Bent
I've a divorce.
Of my acquaintances I fear,
I must confess a chap to be,
Who had the cheek to apply his ear
Where other folk apply a key,
And so I was privileged to hear
The following soliloquy:—
"O sing the joy of a prof. set free
From the prickly chair of Anatomee.
For thirty years I have soaked my brain
In the same vile stuff—but never again!
And when I returned from each long vac.,
My Anatomy ever refused to come back,
And I had to sit down and cram and cram,
—But I had my revenge at the next exam!
Now the sooner it goes, the more happy I am.
And beneath my breath, I admit it's true,
That it's wearing on the conscious, too;
How many tissues I've made to serve
For the intercosto-humeral nerve;
Tendon and muscle, fascia too,
Aponeurosis—they've had to do.
With a mental wink I have palmed them off;
But now—oh, now, I am free to cough
At the sins of the new Anatomy prof."[89]

Allen seemed to be generally well-liked and respected by his students. Personality and competence aside, there were a number of good reasons why a medical professor would gain considerable respect. Allen was a leader of the profession to which the students aspired to join. He was also very much in charge of all aspects of the School. He was, as previously noted, ex-officio (although very much a hands-on) chairman of the Medical Students' Society for much of the nineteenth century. As a product of the School and the first native-born professor at the University, he must have been almost above criticism. Perhaps he was somewhat parochial. Richard Berry wrote that he 'was an extremely difficult man to work with. He had little or no experience of any medical school outside Australia, and for him Melbourne was almost his own creation.'[90] There was certainly plenty of praise heaped on Allen by the medical students. As 'L' wrote of him in 'Professorial Sonnets':

Hail, native-born, of Alma Mater's brood
The first to climb the professorial chair!
Well pleased is she to leave to Allen's care
The budding medicals. From suckers rude
He grows those flowers that – with gold guineas brewed –
Distil a healing balm. His virtue rare
Oft warns his men free-visions to beware
With zeal for learning and mankind imbued.

Not precept only, but example gives
To cherish well the humanizing power
Of cultured mind. For Alma Mater's smile
He's grateful proved. Through him her influence lives
In wider sphere, fast earner for her dower
The longed-for praise of her old Mother-Isle.[91]

The student magazines regularly published a significant volume of Allen's somewhat saccharine and highly patriotic poetry:

When shadows fall,
And cares surround,
To you I turn,
Ye gum-trees tall,
And lessons sound,
From you I learn.[92]

And yet there must have been a good deal of sincerity in their expressions of respect, because the medical students were not averse to publicly criticising the highest of university dignitaries.[93] For example, the medical students would assemble outside Commencement Ceremonies to express their approval or disapproval of selected members of the departing academic procession: 'cheers were ... given for the favourite dons and professors'.[94] In 1886 according to *Speculum*, the Medical School suffered due to 'the utter incapacity of the Council for the management of affairs' because the university councillors were considered to be beneath 'every thinking man in the community' and the Chancellor was 'very petty and irritable'.[95] Although Allen escaped such direct criticism there were complaints about the teaching of Anatomy. This was not the case for Pathology—both subjects being taught by him. In an article written by 'reformer' in 1889, it was rhetorically asked:

> Has anyone ever acquired any information from an anatomy lecture *per se*? or where is the student who can lay his hand on his breast and truthfully declare he has kept wide awake through one-third of the course?[96]

The author went on to argue that Anatomy lectures were dull and of little benefit to the medical student:

> in a large hall containing from fifty to upwards of a hundred students, we have a Professor with perhaps a carpel bone, or one of the small bones of the skull, and on these he delivers a very learned disquisition, while his audiences are for the most part dozing contentedly, or endeavouring with praiseworthy energy to distinguish little protuberances and depressions which it would take Sam Weller's "pair o' patent double million magnifying gas microscopes of extra power" to detect at such distance.[97]

Sleeping during Anatomy lectures also inspired poetry. In 'A Dream at Lecture', 'Barrabool' explained:

> He spoke about "pronation," and I murmured out " ——nation!"
> Of "Interosseous ligaments," till I longed to knock him flat.
> He talked of "joints rotating", and of long-bones supinating,
> Till I slumbered very gently with my feet inside my hat.
> Then I dreamt of pleasant rambles through the tea-tree and the brambles,
> When the week's work had been done—or, at least, considered done,
> But his "ligaments clavicular and surfaces articular"

Intruded on my visions, and completely spoiled my fun.
When he preached on bones cunèiform, I longed to give him chloroform—
As my slumbering cerebrum caused my lower limbs to toss,
Till I landed Smith's innominate, a thing he *did* abominate,
And he growled out maledictions—he *was* so very cross.[98]

In a number of articles anonymous contributors to the student magazine requested that more sessions dissecting under the supervision of extra demonstrators be provided 'instead of the unsatisfactory lectures and demonstrations which are now allowed us'.[99] With this reform 'there would be no calculating how great would be the benefit conferred to our school, from this thoroughness and clearness in anatomical work'.[100] It seems clear that Allen found teaching Anatomy a burden. Criticism, however, must be dispensed carefully. Berry was no great supporter of Allen. He wrote, however, that the truth of the matter was:

> that Allen never had a chance ... when I got there I found Allen engaged in the impossible task of teaching both anatomy and pathology ... Allen got over some of his difficulties by the ingenious method of teaching anatomy in the post-mortem room, hence the forlorn appearance of the dissecting room as I first saw it.[101]

Anatomical accessories – museum

One of the most controversial consequences of medicine adopting 'the surgical point of view' was the need for a regular and bountiful supply of cadavers, as well as pathological and anatomical specimens—all for teaching purposes. On the 22 November 1893, Dr W B Norcott, resident at the Austin Hospital for incurables in the outer suburb of Heidelberg, wrote to Harry Brookes Allen. The purpose of the letter was to explain the provenance of a head and hand Norcott had recently donated to Allen's Pathological and Anatomical Museum in the University, where it had been located after being moved out of the Melbourne Hospital by Allen in 1883.[102] The hand belonged to a seventy-nine-year-old sufferer of rheumatoid arthritis and the head was that of 'Irish Molly', a patient of approximately seventy-three years of age at death.[103] According to Norcott Irish Molly had been so:

> unimaginable at the Geelong Hospital that the authorities sent her to us. She was the most turbulent and troublesome of patients — her powers of denunciation ... and her choice of language unrefined — she was the cause of the death of a respectable old patient and for 2 years was the terror of our [hospital]. I need scarcely tell you that there was a general feeling of relief amongst her fellow patients when she left us.[104]

It seems from Norcott's description that the head, like the hand, was meant to be both a pathological and anatomical specimen. The case of Irish Molly's head well illustrates one important function of medical schools in the second half of the nineteenth century. In the eighteenth and nineteenth centuries there were two schools in relation to the production of anatomical models used for teaching and display. The school of natural anatomy preserved the bodies and body parts either in alcohol or by desiccation, whereas the school of artificial anatomy closely copied anatomic models, either in wax or plaster. Both traditions are represented in the Melbourne collections—specimens such as Irish Molly's head representing the natural method of preservation. Allen also bought a large collection of wax models from the fine French maker Tramond, some of which can be seen in the Allen Pathology and Anatomy Museum in the Medical School today.[105]

In contradiction to all this evidence, Berry's main complaint upon arrival in 1906 was the complete absence of any teaching material: 'not even a skeleton ... no diagrams, models, osteological specimens, dissected parts for study, lantern slides, microscopes, no books.'[106] In 1887 it was pointed out that on the visit of the Governor to the Medical School, all that could be seen in the Anatomy Department were 'some very dry and meagre specimens' carefully arranged throughout the room.[107] As this was not the case, what had happened to all the material?

In fact the collection of anatomical and pathological specimens predated the foundation of the Medical School. In 1859 Dr Shearman Ralph, subsequently and for a long time President of the Microscopical Society of Victoria, was appointed Collector of Anatomical, Physiological, Pathological and Botanical specimens. For this task, he conducted dissections at the Melbourne Hospital.[108] In September 1898, just over seven years before Richard Berry arrived to take up the post of Professor of Anatomy, the Melbourne collection was described thus:

Professor Allen's Anatomy Class 1890. Notice the Tramond skull on table – see second page of the colour section of this book. Courtesy: The University of Melbourne Archives.

The museum occupies a room 60 x 40 with a wide gallery all round, the gallery being devoted to specimens of General Pathology, placed in wall-cases, while the ground floor has long cases down the centre for anatomical models and specimens, and cases in bays under the gallery containing the collection illustrating Special Pathology. A separate cabinet is provided for calculi. There are about 5,000 specimens in the museum, some being the remains of those brought from England by Professor Halford, others gleaned from various sources in the olden days, but the mass of the collection has been brought together by Professor Allen, in his capacity as pathologist to the Melbourne Hospital.[109]

Before Berry's arrival, Allen had clearly taken all the specimens and models for his new Department of Pathology. In his new Department, the previously named Pathology and Anatomy Museum dropped Anatomy from its title. Berry was going to have to start from scratch. It was perhaps a measure of Allen's anxiety about the new Professor of Anatomy that he made no effort to contact him immediately after his arrival. Berry related: 'one morning, about a week after my arrival, he burst into my room with the startling and original remark, "You're Berry. I'm Allen."'[110] It would seem that Allen did make some later compromises concerning the Museum collection.[111]

Another indication of the poverty of anatomical teaching in Melbourne at the end of the nineteenth century was the freely discussed comparisons between Melbourne, and both the University of London and medical schools in Edinburgh.

Melbourne versus Scotland

One distinctive characteristic of the Melbourne medical course was its very high failure rate. In the first decades of the Medical School failure rates hovered around fifty per cent of the first year intake. Medical students often complained about the workload. Sometimes they blamed the inflexible character of those who administered and taught the course. On 26 December 1910 an ex-medical student, Edward Holder, wrote to the Dean:

> Dear Professor Allen,
>
> As one who attended your lectures in anatomy for four years I write to ask a favour of your influence in obtaining work ... on attending your lecture room for oral examination you told me it was unnecessary to examine me as my paper was not sufficiently well done. I felt since, that I might have shown knowledge sufficient for a *pass* had I been examined then, as I had been studying the subject from Ellis & Gray night and day previously with the bones etc ... but you "failed" me and I then felt it was all over with my prospects in Melbourne, at any rate I went home almost broken-hearted and told my mother (who is one of the earliest masseuses in Melbourne ... and practically a widow with four children to bring up) ... I tried to do teaching but it did not suit my temperament.[112]

It is not exactly clear how one could be 'practically a widow' and Allen's response has not survived, but Mr Holder was not alone in his misery. Many who failed then sailed to Scotland or London, tantalising their Melbourne contemporaries with rose tinted descriptions of the medical courses back 'home'. Although frequently praising their School, the student body in Melbourne were clearly discomforted by this foreign competition. In the December 1884 editorial of *Speculum*, it was announced that a printed comparison of the Melbourne and London courses had been circulating in the Medical School and reported:

> [It is] a very severe test that study pursued at a Medical School ... in a young country, such as ours, should be placed side by side with that of a University of London ... Improvements ... are made annually, but that our system should be placed on the same footing as that of the London University, is a scheme which can be regarded only as wild and impracticable.[113]

Indicative of their major cause for concern, in 1887, the Melbourne students were furious at what they perceived as bias in favour of foreign trained doctors for choice medical positions in Melbourne:

> Why are medical men with foreign papers chosen? Simply because it is the custom of lay committees to believe that a foreign qualification carries with it greater experience than a Colonial one. Supposing this were true—and we can never admit that—it can only be because foreign hospitals are reserved for their own men, a lesson we might well profit by.[114]

One of the often repeated benefits of studying in the United Kingdom was the superiority of the teaching of Anatomy. As a past Melbourne student 'S. J. R' wrote from Edinburgh:

> I have been going over my anatomy again under D. R. Symington, his preparations and specimens are truly works of art. He has all parts of the body beautifully dissected and fixed in glass cases around the dissecting room, to keep the structures in position, the dissections are embedded in wax and a layer of spirit all over, helps to preserve them. If the climate would allow it, specimens of this style would be a great help to Melbourne students. His transverse sections of the body also show well the relations of the organs

> to one another. He greases the body and then cuts it into transverse sections at intervals of about an inch. As you [would] imagine the study of these is far better than looking at plates or reading books.[115]

Many others repeated the same benefits.[116] An ex-student 'old friend "Dinnie"' wrote:

> I like the course of study here much better than that of Melbourne; the teaching appears to me to be infinitely better, and it is also much more practical. Everything is so well illustrated that the dullest of students can comprehend. The arrangements for studying anatomy are very perfect. There are mounted specimens of the body in every stage of dissection, and all structures are numbered and placed with the corresponding names in a table above the dissection. In addition to the ordinary lectures on anatomy, there are demonstrations every afternoon by Professor Sir William Turner, or Hepburn, the senior demonstrator ... If you wish to score you have to learn your anatomy from the dissected body and not cram it out of books.[117]

Another important factor in favour of the medical courses in the United Kingdom was that they were surprisingly inexpensive when compared to the Melbourne course. In *Speculum* in July 1887 'Wanderers' pointed out:

> a long course at Melbourne means the expenditure of a very considerable sum of money; indeed, in most cases, more than double the cost of the same thing at the English schools; so this is a very good reason why many of those who either through indiscretion or accident have missed their exams, should prefer to go to other lands to finish their training.[118]

In 1890 it was reported that there were over two hundred Australian students at Edinburgh alone, of whom three quarters were medicals.[119] This corresponded with a decrease in the numbers enrolling in the Melbourne course.[120] Even with the inclusion of travel and accommodation, a degree in the United Kingdom was still cheaper. An audit in 1892 showed that the total cost of a medical degree to a Victorian (allowing for one failed year) was £900, whereas in the United Kingdom it was £650 for a five-year course, or £500 for four years.[121] What was so desirable then about studying medicine in Melbourne?

Anatomy Museum 1907.
Courtesy: Medical History Museum, The University of Melbourne.

It is important to note that many of the reforms brought about by Berry, who had come from Edinburgh, meant that Melbourne students were to be given a course and learning aids similar to those Australians taking the United Kingdom courses, thus eventually nullifying the criticism that: 'the Melbourne Medical Course is at present the most expensive in the world, and the result is that many of our men seek the opposite extreme, Edinburgh, where the fees are smallest and the teaching best'.[122] In a lecture in Melbourne in 1902, J T Wilson, the Professor of Anatomy from Sydney (later Cambridge), probably best summed up the dilemma for the Melbourne course:

> By far the greatest obstacle in our progress ... is the very great insufficiency of staff and equipment in every one of our Australian Universities. It may be that this is the unavoidable result of the comparative poverty of this part of the world ... We are in duty bound to make plain how real our inferiority is. If we are to progress beyond our present third-rate status, far more ample means must be forthcoming than are at present even contemplated.[123]

Both Berry and his successor, Frederic Wood Jones, would repeat this mantra at a later date.

Throughout the second half of the nineteenth century, the Melbourne Medical School had aimed high with a unique curriculum of five years of study, including three in Anatomy. But, until the government provided the finance to set up a Chair in Anatomy, it was clearly not possible to give the students the best possible Anatomy course. It also seems as though Allen was not as committed to the teaching of Anatomy as he was Pathology. This did not mean that the School had failed to produce eminent doctors nor a vibrant medical community. Harry Allen, George Syme, Fred Bird, Jean Greig, Margaret Whyte, Felix Meyer, James Barrett and John Springthorpe among others, are ample evidence of the quality of a Melbourne medical education. By 1901 Melbourne graduates were well on their way to dominating medical life in Victoria. Of the total number of registered practitioners, one hundred and forty-seven had trained in Melbourne and four hundred and fifty-four in Great Britain. Ten years on, the numbers were even.[124]

But for improvements in the teaching of Anatomy, the Medical School had to wait for the arrival of Richard Berry in 1906. Even with shortcomings in the teaching of the subject in the nineteenth century, the intensive study of anatomy legitimised the university trained medical practitioners above all others, by providing the professional with the learning that was essential in establishing status. For the nascent specialisation of surgery, it provided the necessary practical training that enabled previously unheard of operations to be carried out. The growing status of the Melbourne medical community was based on the study of Anatomy. Clearly Anatomy was the central edifice of the triumphant University Medical School between 1880 and 1905.

CHAPTER FOUR

Dissection and the rise of the medical profession: 'being hacked to pieces' or 'foolishly sentimental'?

Sweet is the lore which Nature brings;
Our meddling intellect
Misshapes the beauteous form of things
We murder to dissect.

WILLIAM WORDSWORTH, BRISTOL, 1789

THE STORY OF the supply of corpses, or cadavers as they were known in the medical schools, for anatomical dissection was one of the great scandals of the nineteenth century, especially achieving public notoriety in the United Kingdom. Mary Shelley's novel *Frankenstein* (1818), expressed widespread contemporary fears about cadavers and dissection. Even as late as 1933 public aversion to dissection was still evident in Victoria, as the fourth Professor of Anatomy at the University of Melbourne, Frederic Wood Jones, told the Medico-Legal Society of Victoria:

> The subject of the dissection of the human body cannot be approached without some understanding of the prejudice which inevitably clings to it. We must admit that, there is, interwoven in the whole story, the idea of the sanctity of the human body, the dislike for its mutilation and the demand that it should find sepulchre intact.[1]

This need for cadavers had come about as the result of a dramatic revolution in the study of medicine. The Melbourne Medical School was not to be untouched by controversy regarding the supply of cadavers as it had been established right in the middle of the medical revolution. A steady and abundant supply was essential to the operation of the curriculum. In the first one hundred and fifty years of the Medical School perhaps as many as 15,000 Victorians have provided their bodies for dissection. For most of its first fifty years bodies had been obtained as a matter of course, without the consent of the individuals, and in the face of considerable public concern.[2]

The dominant medical tradition in Western thought until the founding of the Melbourne School was Hippocratic or Galenist humoralism.[3] Health in the body was seen to be the balancing of the four humours—blood, phlegm, bile and black bile (or melancholy). For example, the discharge of blood during menstruation was viewed as an excess of blood, and so bleeding was the obvious therapy. Before the nineteenth century, therefore, a detailed study of Anatomy was not an important part of the university training of a physician because disease was seen to be the product of humoral imbalance and was not seen to reside in particular areas or organs. A plentiful supply of cadavers for dissection was not necessary. Galen, the Roman who dominated the understanding of medicine for over a thousand years, had never dissected a human body.

While dissection of a dead body was undertaken from the Renaissance as part of the education of both physicians and surgeons, to observe or partake of a human dissection was still a relatively rare experience until the eighteenth century. Quite apart from everything else, the church generally discouraged the practice. This meant that dissections (when they took place) were conducted as important events at which not only medical students, but also dignitaries would attend. In the sixteenth century only fifteen to twenty bodies were available for dissection in Paris every year, and in smaller centres it was perhaps only one per annum. Among other events the Revolution in France in 1789 would change all that. Such were the large number of deaths in hospitals throughout the early nineteenth century, as the hospital system was reorganised and made the centre of healing, that hospitals became laboratories of morbid anatomy with abundant research material. Alongside this dramatic increase in the ready availability of cadavers in France, there was a growing recognition that pathological organs and tissues were the centres of disease, not humoral imbalance. As this new understanding of disease was based on anatomical studies, a demand was created for anatomical education.

At the end of the eighteenth and beginning of the nineteenth centuries in the United Kingdom, anatomical dissection was impeded by the legal necessity for anatomists to gain permission from the next of kin for the use of available cadavers, except in the case of hanged criminals. This was difficult because the bodies that were freely available were usually those of itinerants with no known family. This led to the formation of the 'resurrection men' who were illegal grave robbers that provided the necessary numbers of corpses for Anatomy classes. In 1829 William Burke and William Hare were discovered to have provided sixteen suspicious corpses to the anatomist Robert Knox. They had eliminated the need for the middleman, drugged selected individuals and suffocated them, before selling them in prime condition for the purpose of dissection. The notoriety of the case hastened the introduction of the *Anatomy Act 1832*, which allowed for the use of the bodies of unclaimed paupers who had died in the workhouse. What had once been an extra punishment for the criminal (being dissected after death) now became punishment for the poor and itinerant. Hare turned King's evidence at the 1829 trial and Burke was executed, publicly dissected, and strips of his tanned skin were sold to the public.[4] The *Act* greatly worried many of the poor who saw it as a conspiracy by the rich (represented by the anatomists) against them. At various times, for example, during the cholera riots in 1832, crowds of the poor attacked doctors because they believed they had deliberately introduced the disease to provide corpses for the anatomy schools.[5]

From 1832 the first step required to set up a medical school, in the English-speaking world at least, was the need for legislation sanctioning dissection. This was the case in Victoria. On 4 March 1862, Thomas Turner a'Beckett introduced a bill into the Legislative Council 'to regulate Schools of Anatomy'.[6] Thomas a'Beckett noted that as the University of Melbourne was about to introduce a medical degree it was therefore necessary to legalise dissection with an Act that 'was a transcript of the English Bill, with the necessary alterations to adapt it to the circumstances of the colony.'[7] Three weeks later, in the second reading debate, he told the Council:

> It was a misdemeanour at common law to be in possession of a deceased person's body. Of course it would be an outrage of common decency for a man to have anything of the kind; but they all knew what crimes were formerly perpetrated in obtaining possession of dead bodies, for the purpose of anatomical investigation; and how it was found necessary ultimately to legalise what, at first sight, appeared to be a violation of their natural feelings.[8]

The Bill differed in one important matter from the *Act* in the United Kingdom. While executed criminals were not to suffer the further punishment of post-mortem dissection at the University, the *Act* provided for no designated sources of cadavers. As the *Argus* editorial of 7 August 1869 put it:

> it is a deficiency in the Victorian *Anatomy Act* that it does not specify from what sources bodies are to be obtained. It is not too much to say that if bodies cannot be procured from our public institutions, the practice of "body-snatching" will be revived.[9]

It was up to the professor to try and obtain suitable material from wherever possible. At first, and for much of the nineteenth century, unclaimed corpses from the Melbourne Hospital constituted the main supply of cadavers. Prisons and the Lunatic Asylum at Kew were of little use. As Halford complained to the committee of the Benevolent Asylum in 1873, 'in prisons and lunatic asylums *post mortem* examinations are made, which render the bodies almost useless to us.'[10]

Were the crimes (specifically grave robbing) referred to by a'Beckett in the Victorian Parliament in 1862 those of Burke and Hare, or some colonial scandal? Certainly in June 1873, the *Australian Medical Journal* predicted that with the dearth of corpses available for dissection at the Melbourne Medical School and the trouble at the Asylum, which had been 'the occasion for much foolishly sentimental comment in some of the papers ... it will again be the practice to watch graveyards'.[11] What *was* the trouble at the Asylum that provoked foolish sentimentality in 1873? Had grave robbing become a problem?

From the earliest anatomy classes in the University, Halford had difficulties in procuring corpses for dissection. It is difficult to ascertain

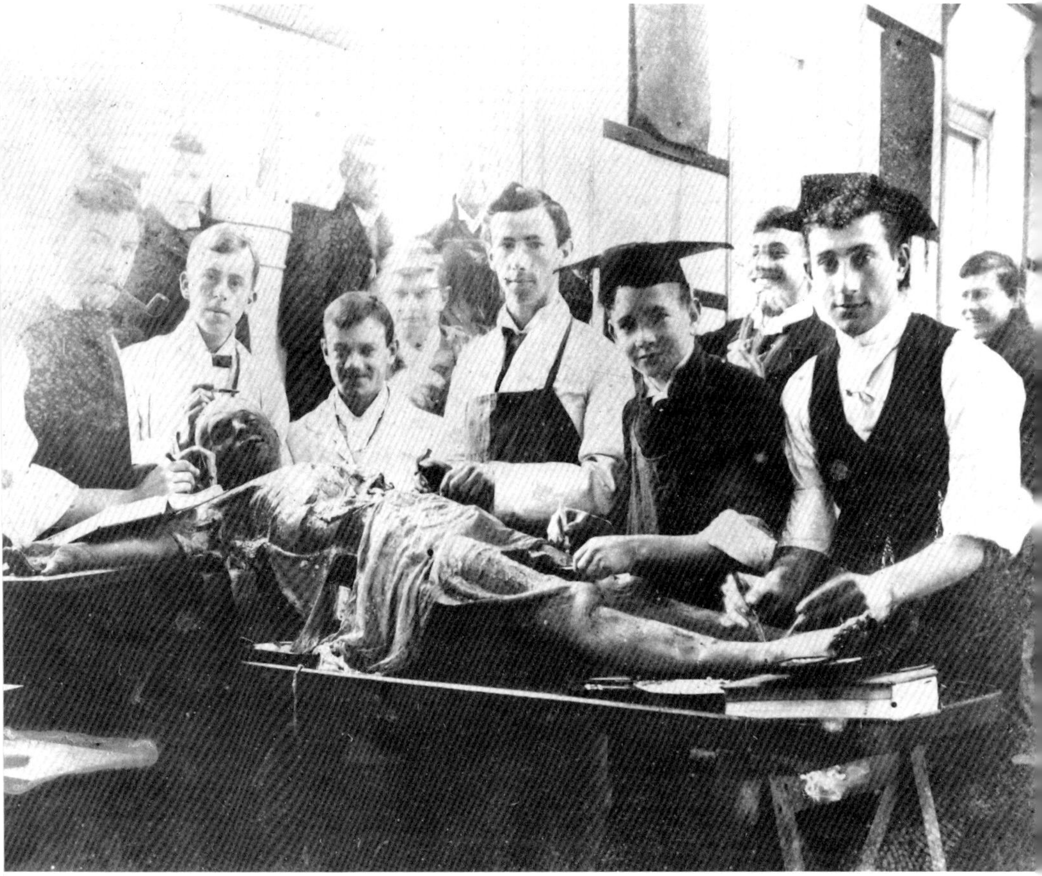

Melbourne students dissecting, *circa* 1894. The tall student in the centre of the photograph is R J Bull, later Lecturer in Bacteriology. Courtesy: Medical History Museum, The University of Melbourne.

the earliest period of shortage. An 1869 account mentions early unsuccessful attempts by Halford to persuade the Committee of the North Melbourne Benevolent Asylum to supply the Medical School with unclaimed bodies. A letter to the *Argus* in 1869 pointed out that Halford had made earlier requests to the Committee due to a 'want of dead bodies for anatomical examination experienced at the University of Melbourne'.[12] These first requests had been rejected, according to the author, because of the 'erroneous notions ... that students were allowed to remove portions of bodies from the dissecting room ... [and] that the bodies of prisoners were removed to the Department of Anatomy.'[13] Halford repeated his

> the human body, for that he knows it. Probably it is an inspiration, an intuition; it is innate, or perhaps it is to be learned by books.[25]

Yet again, in the first decade after its foundation, the Medical School had found itself embroiled in public controversy due to issues raised through the medium of the study of Anatomy.

The supply of cadavers and the growth of the Medical School

A steady stream of corpses was to be the essential ingredient for a successful and growing Medical School. In 1872 there were merely twenty students dissecting, and Halford had claimed that without the supplement of eight bodies per annum from the Benevolent Asylum, the medical course was in jeopardy.[26] It is not hard to imagine the crisis concerning cadavers when the numbers began to rise rapidly later in the 1870s into the 1880s. In 1883 there were one hundred students dissecting;[27] by the time of Berry's arrival in 1906, the annual number had almost doubled.

In 1884, desperate to get more bodies, Halford petitioned the University Council to gain from the government an amendment to the *Medical Practitioners Statute* of 1865 which specified that 'a certificate of the interment of every body forwarded to a School of Anatomy shall be sent to the Inspector of Anatomy within six weeks of the first reception of such [a] body.'[28] Halford wanted the period changed to six months because 'bodies can be collected and carefully preserved while the classes are not in session; so that, at the commencement of the dissecting term, no delay shall be experienced. No difficulty is found in properly preserving the said bodies.'[29] This was essentially achieved, as corpses were routinely kept in the Anatomy Department for this period by the end of the nineteenth century.[30] The dissecting term operated from May until October, so as to delay the decomposition of the bodies by taking advantage of the colder weather. Halford now wanted to collect summer bodies. There is no direct evidence of what new method was proposed to preserve them. In 1892 in a poem concerning a dream of being dissected, 'Barrabool' wrote: 'When a spectral overseer held his spirit lamp too near, to my spirit saturated corpse, which burnt up with a flash.'[31] Probably alcohol, which had been used for at least a century to preserve pathology specimens, was part of the formula.

While Halford's efforts may have alleviated the problem he had provided no concrete solution. And 1887 proved to be a particularly disastrous year. This followed closely on from the tumultuous year of 1885 when nine of ten fifth year students failed. This was the 'The Fifth Year Fiasco' so called by *Speculum*.[32] Students commenting on the visit of the Governor, Sir Henry Loch, to the dissecting room on 13 September 1887 complained:

> Had his Excellency personally inspected the room and spied out its nakedness, he would have concluded that, out of consideration for feelings of visitors, the subjects had been hidden away in the cellars, for he certainly never could have believed that this great and flourishing Medical School of ours actually had no bodies for its students to dissect.[33]

Although by this stage it was possible to obtain bodies in the summer and preserve them, the 'fact remains, however, that the supply of bodies has proved quite unequal to the demand; and it is the duty of the managers of the institution to which we pay our fees, to provide some means of securing for students, material for the work, which ... must be done.'[34] One method used to solve the problem was called 'double-banking' which meant having two students work on one part of the body. The students voiced their opposition to this practice loudly and unendingly throughout the second half of the nineteenth century, complaining that the University was breaking its contract by accepting fees for dissection without providing adequate cadavers: 'for years no student has received the amount of a subject that his £4 4s entitles him to, and this year the situation is worse than ever. Surely if only two parts, double-handed at that, can be given to each student, it is not honest to charge for a whole body to himself.'[35] It was argued that the practice of conducting post-mortems at lunatic asylums was a waste of time because it denied material for dissection at the Medical School. Such was the shortage that: 'it is a matter of history that once upon a time in older countries the resurrectionist was a flourishing tradesman. It is not desirable that history should have to repeat itself.'[36] Fear mongering again!

The shortage meant that in 1887 there were inadequate opportunities even for demonstrations in the lecture theatre. *Speculum* complained that 'the word "prosector" [senior student chosen to be a demonstrator and supervisor] conveys but little meaning to many students who are no longer juniors.'[37] Students also cried out for the teaching aides available in Great Britain but absent from the Melbourne course:

> is it of any use to wonder if we shall some day have prepared dissections and frozen sections of the human frame by an occasional observation of which students might impress on their memories a much more accurate knowledge of the arrangements of structures and organs, that can ever be gained by studying plates or descriptions.[38]

All of these needs were to be supplied by Richard Berry after 1906.

The criticisms clearly indicate that the course seemed to be at risk, especially as cheaper courses in the United Kingdom beckoned. As the Medical School in Melbourne comprised half the students attending the University, this was a serious problem indeed. In correspondence to *Speculum* in 1887 the depth of dissatisfaction is unbounded:

> Sir, I write to you on behalf of the other students as well as myself, who are, or rather ought to be dissecting at the moment ... Few of the students in any of the three dissecting years have, during the past three terms, got further than their second part, although we are now in the middle of the last term.[39]

Although the correspondent had paid his fees for dissection he had not received his proper due. There were three questions for which he demanded an answer: '1. Does the course include four parts—an entire body?, 2. Should we get back fees?, and 3. Are we expected to know for examinations those parts which many of us have not seen?' He went on to argue that the examination on parts of the body they had never dissected:

> would seem to be a very great injustice ... It is quite possible that a student who has had only two parts this year, may find that most of the questions in the paper set from that half of the body which he has not done ... His practical knowledge of Anatomy at this rate will be far short of what it ought to be, a fact which he will never cease to deplore when he is practicing surgery.[40]

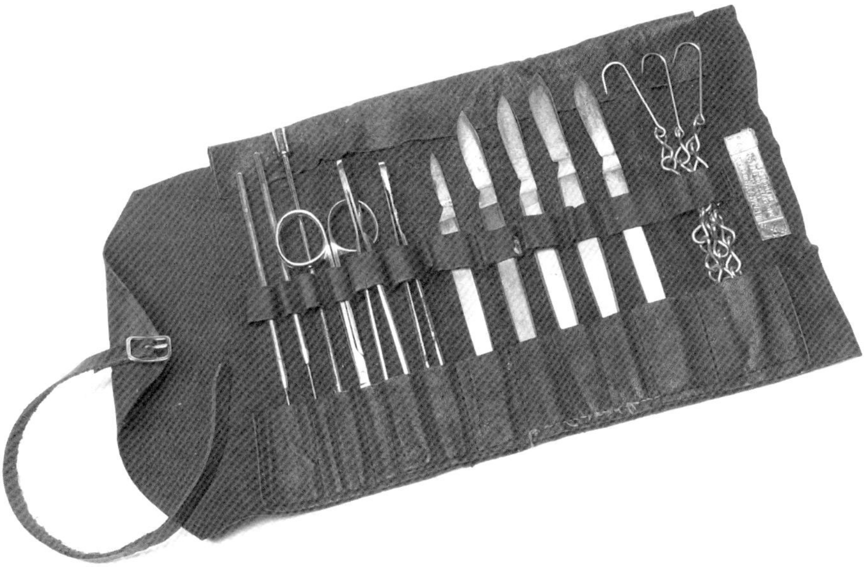

Dissecting kit in use in the early twentieth century. Courtesy: Medical History Museum, The University of Melbourne. Photograph courtesy: Michelle Gough.

No doubt the patient being operated on would deplore it even more. What solution was possible? It was back to the Benevolent Asylum in North Melbourne:

> The Benevolent Asylum surely could give us a few bodies. The patients there I hear, are all asked before they die, whether they wish to be dissected or not. I suppose the answer is invariably—No! as we never see any of their bodies in the dissecting room. This seems a very curious proceeding. I should rather think that the people who live on the bounty of the government ought to be pleased at being able to make some return to the public who have kept them for so long.[41]

In fact the inmates were not kept by the public purse and the committee representing those who did finance the Asylum had refused Halford's request for bodies. Operative patients, as well as the destitute, had good reason to be wary. In 1888 *Speculum* reported the 'very unpleasant results'[42] in the operative surgery examination the year before due to the lack of cadavers. Tharp Girdlestone had rushed through the final surgery examination for twenty-three final year students in little over an hour and just before the graduation ceremony because no cadaver had been available earlier.[43]

By 1889 the position had improved, though double-banking was still common. It was stated that it was hoped that solutions were at hand 'with pickling permitted, and the supplies from the country hospitals coming in.'[44] The problem was chronic and a deputation to the Dean, Harry Brookes Allen, had been discouraging. Allen had been in contact with every charitable institution in Victoria with little success:

> One in particular, which is supported largely by a government grant, forwarded the childish plea that were they to accede to the wish of Professor Allen it would tend to make the institution unpopular. When we find a committee who urge such an excuse as this, what hope can be expected of a remedy? It may be the individuals complained of have in view the time when they themselves will be eligible for a vacant allotment at the public cost.[45]

The supply of bodies still came almost exclusively from the Melbourne Hospital and the majority later in the year, at the end of winter and just before the final examinations, forcing the students to cram. *Speculum* proposed a new system whereby an Inspector of Morbid Anatomy should be appointed and paid a per cadaver rate, thus encouraging diligent collection and the offer of financial rewards to the institutions: 'His income would be made proportional to the supply, and as we could find room for close on one hundred bodies, he would be zealous in his efforts for our benefit.'[46] The editorial also repeated the mantra of the university trained, scientific medical practitioner:

> In these days when science demands a thorough knowledge of every branch, no obstacle should be allowed to stand in the way of any design calculated to further the good of humanity. The government also should be brought to understand that if they wish the medical men of their University to rank with those of other nations, technical knowledge in this as well as other things is

> required. And as this technical knowledge is only acquired by the aid of morbid anatomy, we should use every legitimate ends to further that end.[47]

Failing to pass was also important: 'The disastrous results of the recent February exams, in third and fourth year subjects, led us to emphasise the same point in a letter to the Council, and we have been informed (semi-officially) that the University will begrudge no expense that may be incurred.'[48]

The measures put in place by Allen seemed to have worked, because in 1890 there was a 'great rush of corpses'[49] and double-banking was abandoned, allowing each student to work on his or her own corpse. This continued to be the case until 1894.[50] In that year *Speculum* reported that a 'prevailing depression has invaded the dissecting-room. Four bodies for one hundred men. Operative surgery in similar condition ... Wanted to know – How to learn Practical Anatomy without dissecting?'[51] This continued until 1897.[52] In 1898 the complaints stop, and it seems it must have been the introduction of the preservative formaldehyde that led to the sudden improvement in the supply of cadavers.[53] It may be a coincidence that in the same year the first mention is made of the acquisition of an Aboriginal corpse for dissection.[54]

Women medical students

In 1887 women were finally admitted into medicine—seven years after they had been admitted to the university and twenty-five years after the foundation of the Medical School.[55] They received a mixed reception. 'Rotomahana' in his poem 'To the Ladies' wrote:

> O maidens fair, of Austral's sunny shores,
> Ye stars of wisdom, that my soul adores.
> What new-born fad is this that fires your mind
> To seek for knowledge of another kind
> Than that, which hitherto has moved the hearts
> Aspiring to the dignity of Arts?[56]

From the beginning there were complaints from some of the males that the women sat in the front row in lectures and failed to withdraw, as the nurses did, from *delicate* operations. How would it be possible to learn surgery if the women were unable to observe operations? As one woman replied:

> we invariably absent ourselves, both in the wards and in the theatre, from operations which we will never meet with in practice, and which lie wholly outside our future sphere of work. Is there no indelicacy in male students attending special operations on women?[57]

They also caused animosity among their opponents due to the extra pressure they placed on limited clinical teaching resources. Perhaps the most trenchant criticism came about because for many years they insisted on having a separate dissecting room, unlike the women in Sydney. This was during the extended period when corpses were as rare as hen's teeth. It seems as though the women's dissecting room suffered from the general dearth of dissecting material.[58] They finally joined the males for dissecting after the arrival of Richard Berry in 1906.[59]

It is clear that one perennial and often insurmountable problem affecting the teaching of Anatomy was the supply of corpses. Just a few years before Richard Berry's arrival, the problem ceased. Berry's hard work in providing other teaching aides, such as preserved sections of the body, must also have greatly improved student's access to the anatomy of the body. Apart from Allen's moderate competence and his lack of interest in teaching anatomy, another problem affecting its effective teaching was the lack of adequate supervision.

Supervision

In 1900 *Alma Mater* complained about the absence of supervision in the dissecting room:

> Medical students have been accused of nearly all the ills that flesh is heir to, but this mystery of Carlton is beyond a joke. It passes comprehension how even a pranksome student could appropriate — or misappropriate — the gruesome specimen referred to (to whit 'the leg of a male adult of small stature') without the proper knowledge of the authorities. Surely some more adequate system of superintendence is badly needed in the Dissecting Rooms.[60]

I have already mentioned the problem with having just one busy surgeon demonstrating during the week, when perhaps one hundred students were spending extensive periods of time at the dissecting tables. Clearly senior students (prosectors) were required to shoulder the burden of the supervision. In the year of Allen's absence on leave, the students published their own handbook of dissecting room etiquette because 'during the years in which the Medical School has been in existence a number of unwritten rules have come to be tacitly observed in the dissecting room'.[61] In the dissectors' vade mecum (manual), it was pointed out that only certain rituals and costumes were acceptable:

> It is considered somewhat old-fashioned to wear an old coat and long apron in the room now-a-days. Fourth and second-year men, especially, who have to deal with surgical cases at the hospital, seldom go to the trouble of changing their coats: demi-sleeves of some blue material slipped over their ordinary coats are all that is considered necessary.[62]

The students recognised that 'a few bacteria are from time to time carried from the room to the wards', but they believed that 'Listerism' [The use of carbolic spray to sterilise the hospital] is so scrupulously carried out in the M. H. [Melbourne Hospital] that they never obtain admission to wounds'.[63] Dress was particularly important:

> college caps look jaunty, and are therefore much affected by college meds. Mortar-boards have the advantage that their tassels mop up any superfluous moisture on the part being dissected. Pot hats are extremely comfortable and "becoming", and have the advantage of preventing the man opposite from approaching his part.[64]

It was possible, according to the manual, to establish those of seniority because of the confident gait they affected. They must also confirm their superiority, when arriving, by wandering around and criticising the juniors' work because 'it impresses unsophisticated second-year men, who may mistake the senior for a demonstrator ... it is considered *infra dig* to replace coverings on a part which you have examined—a porter is kept chained up on the premises for this purpose'.[65] The prohibitions on smoking and being noisy are rules 'more honoured in the breach than the

Early women medical students had their own dissecting room.
Courtesy: Medical History Museum, The University of Melbourne.

observance; besides, infringement of these minor rules gives a devil-may-care air of disregard for the authorities, which suits well the conventional idea of a medical student.'[66] As far as seeking entertainment when enduring the long hours of dissecting, the student is encouraged to:

> practice aiming chunks of semi-putrid muscle across the room at their bosom friends. ... at first you will probably miss ... and hit some inoffensive med. Unexpectedly you may be so unfortunate as to hit an aggressive med. This chance adds an element of danger to the pastime which increases its zest. Your friend will probably retaliate, according to ancient custom, by discharging some putrescent internal organ at you, which will equally probably miss you and "land" someone else. The fun is thus kept up *ad libitum*.[67]

Furthermore, in the past students would read up on the part they were dissecting the night before but 'this practice is out of vogue.'[68] Finally, the student, 'on leaving the room for the day', should pocket a femur or pelvis from the skeletons 'which the authorities have considerately supplied to enable students to complete their own sets of bones.'[69] Richard Berry was appalled by this entrenched behaviour among the students. He wrote:

> Had these bodies been those of an oxen instead of humans, an old time butcher would have said it was a waste of good rump steak. A grim smile was my only comment on this curious practice which ceased henceforth.[70]

Corpses or cadavers?

Who provided the bodies for dissection? The first surviving recorded cadaver in the Department is that of James Halferty, aged fifty-three, whose last place of abode was the Fitzroy Cricket Club in Brunswick Street. Most likely he was itinerant and he is recorded as being of the Roman Catholic faith. He died on 24 October 1898 in the Melbourne Hospital of heart failure caused by arteriosclerosis. His corpse was received in the Anatomy Department at 8pm on 25 October 1898. The certificate recording his death and burial had been signed at the Melbourne Hospital by Violet Plummer, MB, BS (one of the first elected, female resident medical officers and a graduate of the Medical School), and in the Department upon arrival, by Professor Allen.[71] If any records were kept before 1898 they have disappeared, as the records held by the Department start on 26 October 1898. The 1900 records appear to be missing, but records begin again in mid-1901 and continue to the present.

In 1902, the first full year of records, the Department received fifty-five corpses at the normal cost of £1 each. Cheaper corpses were assigned to operative practice and their condition must have been too poor for dissection classes. Although the North Melbourne Benevolent Asylum had resisted the overtures of the Medical School in the earliest decades, by the turn of the century a good proportion of the corpses were supplied from another benevolent institution for the itinerant, the Immigrants Home in Royal Park. In 1899 twenty-two bodies came from that institution for the homeless; in 1900, seventeen; in 1901, ten and in 1902, nine. A smaller number came from the Victorian Houses for the Aged and Destitute in Parkville and St Kilda (only two in 1902) until they became a major supplier by 1910.[72] In 1902 only thirteen of the fifty-five corpses received came with addresses that included a street number. Twenty-one were supplied with the name of a suburb only, and with these, occasionally the name of a street was given. As some of the younger women gave their address as Little Lonsdale Street, it is highly likely that they were prostitutes. Two resided on ships docked in the port and both men had died from phthisis, the common description at the time for tuberculosis. Seven had no known address. Of the total number more than three-quarters seem to be itinerant or semi-itinerant. The cause of disease frequently point to tragic ends. Seven died of alcoholism and eighteen from tuberculosis—nearly half. The average age of corpses was fifty-five, almost exactly that of male life expectancy in 1902. This was a rather surprising statistic when considering the tenuous nature of their circumstances in life.[73] A student poem from 1894 commemorating the death of the dissecting room porter describes the bodies in terms of their demise:

> When Winter saw the typhoids end,
> Pulmonic troubles did not falter.
> Then, just as "go-betweens," there'd come
> Some "cardiacs" or "fractured bases",
> Chronic kidneys, or likely some
> New fangled "op" to fill his "cases".[74]

Clearly the cadavers dissected by the medical students came from the poorest classes of society.[75] While the larrikins of Carlton, North Melbourne and Fitzroy may well have attacked the processions of medical students, the meds would meet up with them in the dissecting room later and take their revenge.

Class and the rise of the medical profession

I have already mentioned in an earlier chapter the sense of camaraderie that was an essential element in medical training at the University. This was reinforced by the physical separation across the lake, the very full timetable that restricted any activities outside the School, and the common destination of graduates into a unified profession that was, at this time, vigorously establishing an important niche in upper middle class colonial society. The rituals of the dissecting room – including meat fights – would seem to be an essential part of that singularity, according to students. Spending so much time in the presence of dead bodies would also have served the purpose of desensitising the young student, and producing the detached professional that was seen to be the ideal type of practitioner. Like officers on the battlefield, meds attempted to trivialise the cadavers by characterising them as 'other'—not one of them.

This detached attitude would have been aided by their social isolation from the lower classes that provided the bulk of the bodies. A report of a medical student night out at the theatre characterised the audience in another section of the theatre as identifying the meds as superior socially: 'general murmur of "s'help me Tommy they're all toffs, look at their faces." Please note this remark, oh! Physiognomical Dr Moloney.'[76] This allusion to working class physiognomy represented the then commonly held medical view that there was a distinct physical indication of the biological inferiority of the lower classes. *Speculum* characterised Carlton, the home of its working class neighbours, as 'Fairyland':

> Who is there among us that does not love Carlton? Is it not through its cheerful streets that, year after year, we have trudged up from the M. H. [Melbourne Hospital] ... enjoying the enchanting scenery, the bracing atmosphere, the music of the ever-joyous larrikin's plaintive accordion ... Here we see the pleuro-pneumonic cow, also the bony variety ... or the bounding mongrel of Carlton, which has the peculiarity of always being in a state of mange ... stray carcases [of cats] are found lying blistering in the sun, around which the harmonious blow-fly and the giddy bacillus frolic in thousands.[77]

The impoverished resident of Carlton seemed as if from a foreign country:

> Swarms of young Carltonians may be seen and heard at all times of the day and year dancing and singing – dressed in their own peculiar, and at the same time picturesque, garb – to the strains of the soul-stirring, never-wearying hurdy-gurdy or street organ, played in masterly fashion by a swarthy "Irishman".[78]

The Carltonian was a quaint and distant character, a:

> "singular anomaly", the Carlton masher ... who, with thumbs in breeches' pockets, a large mushroom, carefully ironed, hat, "overhanging foliage" on his forehead, and the irresistible bell-bottomed pants ... saunters along, and loiters around the sunny corners of the blue-stones, dreaming love's young dream, or of pints of ale to come, and anon squirting from beneath his teeth the juice of the succulent tobacco.[79]

Familiarity with the dead body was also a privilege reserved for the University medical student. In the April 1899 edition of *Alma Mater*, it was reported that working class youths were visiting the morgue to view the remains of fatal accidents and suicides. In high dudgeon, a medical student wrote:

> With an elementary system of compulsory education [introduced in 1872], of which it is our boast that it has rescued our youth from crass ignorance, it is rather a takedown to find that young girls, in crowds, pass through this dead house and make merry over the mutilated remains of a wretched sister. We cannot remember the slums of any large city, in any part of the civilised world, ever turned out greater streams of heartless humanity than those which surged through the morgue to glare at the head of a poor young girl, who, however giddy, had been cruelly and clumsily done to death.[80]

The apparent insensitivity of the medical school student in the dissecting room, like the officer on the battlefield, created among some the type of neurosis that saw successful professional men and scientists turn to the irrationality of unconscious guilt, pre-Freudian neurosis and spiritualism. John Springthorpe, father of a dynasty of psychoanalysts, active member of the Medical Students' Society, lecturer in Therapeutics and successful President of both Melbourne's Artists' and Bicycling Societies, spent every evening for decades from 1883 recording his daily dreams of his departed wife Annie in extraordinary fashion and detail. Along with the written record of the dreams on each page were poems lamenting her departure; cut-out pictures of pre-Raphaelite maidens with photos of Annie's head substituting that of the maiden in the picture, along with angst-ridden homilies about the absence of God and the possibilities of a form of spiritualist reunion.[81] Other Melbourne medical students in the late nineteenth century perhaps exorcised their guilt in the form of macabre poetry about the activities in the dissecting room.[82] 'Barrabool' among others wrote such poetry:

> And then the Dream-god caught me, and to sights still stranger brought me:
> I was laying on a table in a lofty lighted room,
> With my *corpus* half dissected, and my joints well-nigh bisected,
> And I hadn't got much time to spare before the Crack-o'-Doom.
>
> And all my fellow "stiff-uns" were dancing around like griffins,
> After missing bones and muscles which they once had called their own;
> For St Peter and another had raised a lot of bother, and wouldn't resurrect them if they wanted half a bone.
>
> There was Brown of Booroondarra, who was drowned in Lower Yarra,
> He was sneaking out a femur from John Selby's private hoards,
> While Snookson, from Tasmania, had developed raving mania,
> For he couldn't find his biceps on any of the boards.
>
> Another dear departed was nearly broken-hearted,
> For a second-year dissector had completely *lost his head*.
> But *I* lay still and quiet thro' all the ghastly riot,
> For the muscles of my dorsum to the yard-dog had been fed.
> I tried to laugh, but couldn't, as my facial muscles wouldn't
> (For Simpson, when dissecting, is a bit inclined to hash),
> ...
>
> Then I heard a mighty tolling of a bell, whose sounds were rolling
> As though the very vaults of heaven their harmony would break,
> Then I felt a someone hauling at my gown, and heard him calling,
> "The lecture's done, you idiot—wake up, for heaven's sake!"[83]

Dissecting working-class bodies would have helped the students to distance themselves from the daily reminder provided in the dissecting room of their own mortality.

Creation of a united profession and the Medical School

An important aspect of the development of the University of Melbourne trained medical practitioner as a respectable, professional expert in the community, was the sense of camaraderie forged through the common culture of the Medical School, as well as a singular aspiration to higher social status. Dissection was at the very centre of the course, and it enabled the university trained to distinguish themselves from alternative practitioners, and from those who would become their patients. It was not just money that enabled the Melbourne medical graduate to distinguish himself (and later herself) socially. At this time many of the graduates found themselves practising far from even a large town, and struggling to make a reasonable income.[84] As *Speculum* complained in July 1887 'money is frequently a question among students, who are not *all* wealthy.'[85] The students were, however, due to the expense of fees, not representative of the poorer classes. Medicine was seen to be producing a new scientific and professional elite. As the Melbourne Medical School was slowly but surely taking over the medical profession in Victoria, it was edging toward a pre-eminent position among this new class. As the editorial in *Speculum* ruminated in 1887:

> In this age of cheap jewellery – and especially in a country like this, which has been populated by money-getters – it is only to be expected that there would be many willing to undertake the labour of a course, say of medicine or law for the sake of *kudos*, and the social standing, which are among the prerequisites of the professional man. It cannot be thought of such, that they spend a moment's more labour in study than they find absolutely necessary.[86]

Also, as Professor Allen regularly preached, the medical profession had responsibilities due to their increased status, and this was essentially to maintain Imperial values and culture.[87] Although hard won, the increase in status did not guarantee universal accolades. The student body knew that greater efforts would be required to win over the population: 'There is deep prejudice among the public against medical men. They are said to overcharge and be too grasping, and we are sorry to say that in some instances this is true.'[88] The development of the image of the medical doctor as scientific expert would increase the reputation of the medical profession in Melbourne.[89] The great development of the study of Anatomy after 1906 and under the direction of Richard Berry provided the basis for the confirmation of the status of doctors in Victoria as medical scientists. This status is reflected in the cadaver archives of the 1950s and 60s. By then, the supply of cadavers came from donations and Melbourne's middle classes. In fact there were so many that a large proportion of cadavers were rejected. And donating one's body, by this time, was seen as making a selfless contribution to the triumphant progress of medical science.

The Anatomy School 1910.
Courtesy: The University of Melbourne Archives.

CHAPTER FIVE

Richard Berry and the measuring of the race

Now all of us our wives must take
For Health and Race improvement's sake
The mother's an anatomist
And Cupid's with an oculist!

EDWARD DYSON, *LONE HAND*, 1907[1]

ON THE EVENING of 30 November 1917 shortly after 8pm Richard Berry, Professor of Anatomy at the University of Melbourne, rose to address a public meeting at the North Melbourne Town Hall—one of three predominantly working-class suburbs surrounding the University in Parkville. On the following day, the chaotic scenes that took place at this meeting were described by the *Argus* and *Age* newspapers. The *Argus* chose the headline 'Meeting Broken Up – National Anthem Derided', while the *Age* described it as 'A Disorderly Meeting – Speakers Refused a Hearing'.[2] The chairman had suggested they open the meeting with the singing of the national anthem, and the *Argus* reported:

> the opposition section broke out into shouts of derisive laughter. They remained seated, the men wearing their hats, and endeavouring to drown the voices of the singers by stamping and blowing trumpets.[3]

Following the singing of the anthem, various speakers were forced to stop in the face of incessant interjections. Professor Berry, the *Argus* informed its readers, only managed a few audible sentences before some of the 'rowdy element' went over to the side doors near the stage 'and by kicking at them drowned every word that was said inside the hall', the result being that he was forced to retire from the stage without having delivered his address.[4] The date of this meeting gives the clue to its subject matter: it had been called to lend support for the 'Yes' vote in the second conscription referendum of the First World War.[5]

This episode is revealing and tells us a great deal about Richard Berry. Orphaned shortly after his birth in 1867, Berry came under the care of his grandfather who sent him off to a Dame's school (a Victorian era child-minding facility), and later a private school, both in Lancashire. He then became a shipping clerk in Liverpool until early in 1886 he threw caution to the wind and managed to inveigle himself into the medical course at the University of Edinburgh. His grandfather, thrilled with Richard's ambition, funded the experiment.[6] Coincidently, many failed students from the Melbourne course were to be his compatriots in the premier Scottish Medical School, thus giving the unwitting Berry the first glimpse of what would be a long-term connection with Melbourne. Although he was not able to apply to Oxford or Cambridge (as he was not a public school man) Berry had no regrets, as the reputation of the Edinburgh course at the time was unsurpassed. It also gave him his first taste of cosmopolitan life, and he later wrote that when he commenced his studies:

> The fame of the medical school was world-wide. Students flocked to it from all parts of the world ... English, Scottish, Irish and Welsh ... West Indians, black men from the Sudan, Princes from India – the Thakor Sahib had a special seat reserved for him in all classes – yellow men from China and Japan ... Australians, South Africans and New Zealanders, and even one or two Americans from Buffalo and California – I rubbed shoulders with all of them but spoke to few.[7]

Unlike the positive sentiments expressed in letters sent home from Australian expatriates, Berry was critical of the limited interest the University took in its students, as well as much of the teaching he encountered. He complained that 'apart from providing the class rooms and laboratories and collecting the fees, the University ... did absolutely nothing at all. It was just a gigantic day-school, where students were left to their own devices. It either made you or marred you and many fell by the wayside.'[8]

Berry was taught Anatomy by Sir William Turner, who in turn, had been a student of James Paget, patron of the Melbourne Medical School and Halford's selector for the inaugural Chair. Although somewhat critical of Turner's teaching, Berry must have imbibed Turner's methods and interests because he reproduced them later in Melbourne. Berry's use of numerous teaching aides, such as preserved cross sections through the body, his Darwinian stamp of mind (Turner had advised Darwin on anatomical subjects while remaining uncommitted) as well as his obsession with Anthropology and Craniometry, all meant that Melbourne became a close facsimile of Edinburgh.[9] Berry also developed an Anatomy Museum much like Turner's famous institution in Edinburgh. When wandering as a student through the museum, Berry would have observed the skeleton of William Burke, the murderer who supplied the Scottish anatomist and anthropologist, Robert Knox, with bodies for dissection. Berry's 'The Anatomist's Song', written while he was in Melbourne, pointed out that 'The man ... who Anatomy wishes to shirk, shall in consequence kill all his patients, and so furnish subjects as did Mr Burke.'[10] Burke had been dissected in Edinburgh after being hung. It seems that Anatomy students received a guinea purse fashioned out of a scrotum at the end of their course. And it is believed that William Burke contributed such a gift.

From the moment he arrived in Melbourne in the summer of 1906, Berry threw himself vigorously into life at the University and into the wider community. The *Herald* opined in 1923 that 'Professor Berry has impressed his personality not only on the University but on the community ... a caustic tongue has the Professor'.[11] Others reckoned he had a 'violent temper' and used 'very strong language'.[12] Although Berry made many friends, his prickly nature meant he also made many enemies. As we have seen from his reception in the North Melbourne Town Hall in 1917, it is clear that in his role as public thinker and opinion maker, Berry could arouse strong opposition. He thought little of Allen, Berry characterising their working relationship 'as difficult a combination as can be met'.[13] He also conducted a lifelong feud with the ophthalmologist Dr (later Sir) James Barrett, one of Allen's pupils, who eventually become a Chancellor of the University. It was said of Barrett that he never joined an organisation without aspiring to, and subsequently gaining, leadership.[14] *Speculum* published an anonymous tale illustrating the students' view of Barrett's ambitions:

> A certain prominent Melbourne eye specialist noted for his wide interest in public affairs had departed this life. On entering heaven he noted the very unsatisfactory manner in which celestial affairs were conducted. He immediately suggested to St Peter that he should draw up a workable constitution. A few hours later he returned and handed the result of his labours to the Guardian of the Gate. The Apostle looked at it for a few moments, then: "Yes, this is all right, Doctor, but I am afraid you cannot make God vice-president."[15]

There are numerous amusing tales concerning Barrett's and Berry's mutual detestation.[16] Berry blamed 'that bastard Barrett'[17] for blocking his attempts in the 1920s to move the Melbourne Hospital to the University to create a clinical school of medicine.[18] Another of his colleagues, the Professor of Chemistry, David Orme Masson, infuriated Berry by buying two thousand four hundred votes (an investment of £10 at one penny per vote) for Berry in a competition run by the students to select the ugliest member of the University, thus ensuring Berry's victory.[19] His successor Frederic Wood Jones noted after his arrival that 'the medical men in the town have somehow raised a fund for increasing my pay beyond the University salary. I believe Berry is sore about it, but he had treated some of his colleagues rather oddly, for he was a bit autocratic in his own queer way.'[20] This climate of hostility was to cost Berry dearly toward the end of his stay in Melbourne.

As we shall see his regular and frequently packed public appearances throughout Australia were, unlike his experience in North Melbourne in 1917, almost universally applauded. Berry was to follow in Halford's footsteps, bringing notoriety to the study of Anatomy outside the University. With public influence, however, Berry would far outdistance Halford as he was able to place Anatomy right in the centre of the Australian public consciousness. Rarely has the University played such an influential role in public life as it did during Berry's tenure as Professor of Anatomy. Australian educated thinkers, doctors, educators and virtually the whole of the print media embraced his views.

Berry was aware of his opportunity to contribute mightily to the development of the young colony of Victoria and the nascent Commonwealth of Australia. He wrote that '[a]s there is no aristocracy of birth in these Dominions overseas, aristocracy of intellect takes its place. Consequently a Professor of the University of Melbourne occupied a very special place in the society of my time'.[21] Berry was involved in a number of high profile public campaigns. The most important campaigns were the reform of the teaching of Anatomy, along with the creation of a clinical teaching hospital on the University site (the subject of Chapter Seven), and combating the supposed threat posed by 'mental deficiency' to the health of the Australian race.

Berry was in no doubt about which topic he believed to be the most important. In 1939 and after ten years of service at the Stoke Park Colony for mental deficients in Bristol, England, he wrote of his growing belief in the need for euthanasia as well as sterilisation in order to reduce the incidence of mental deficiency in the community. Reflecting on his belief in the need for such extreme measures, he remarked:

> if this had been all to the subject of mental deficiency, or as I prefer to call it, cerebral insufficiency, I should not have wasted the best years of my life on it, nor should I have spent so much time trying to educate people up to the fact that the danger to civilisation comes not from the segregated idiots and imbeciles in institutions, but from neighbours and fellow voters – so many of whom are quite definitely of the feeble-minded class.[22]

The mentally deficient were the supposedly inferior citizens whose identity, as described by Berry, we will examine later, as well as his plans to reduce their numbers.

National efficiency

In the manner that 'market forces' and 'economic rationalism' swept all economic worldviews before them at the close of the twentieth century, so did the idea of 'national efficiency' a century earlier.[23] As discussed in Chapter Two, Darwinian ideas infiltrated social investigation and it was widely believed that the nations of the world were in a battle for evolutionary supremacy. This was one aspect of what came to be known as social Darwinism.[24] Berry was a prophet of 'national efficiency', claiming that the study of human anatomy was an important tool in prosecuting social reform and it was anatomists who could accurately identify the mentally deficient. Berry argued that the burden he believed that the mentally deficient later placed on the community through the cost of social welfare, lost productivity, social unrest and institutionalisation was severely detrimental to Australia's development. As for finding a solution to the problem, he believed:

> it should be clear to every unprejudiced person that the work is of the utmost social and economic value [and that] ... it will be possible to pick out, with a precision never previously approached or surpassed, the possible inefficients of adult life from the abnormal types of school life.[25]

As Berry's concerns about the malevolent influence of the mentally deficient were allied to wider concerns about the efficiency of society, it followed that in Berry's view, the inefficient had to be recognised and improved or stopped altogether from breeding. Also the climate was right for Berry's views as the degeneration of the racial stock was an underlying fear of most middle-class Melburnians.[26] During the First World War, Berry wrote a regular opinion column in the *Herald* newspaper in which he frequently took over a page to present his arguments promoting the 'Newer Imperialism'; a version of national efficiency.[27] This was an economic and political philosophy that promoted intelligence and efficiency over 'one man one vote' representation. It was an early form of fascism and fashionable among the educated middle class.[28] In these articles Berry argued against the trade unions and the right to strike, and frequently and angrily demanded a reduction in the number of parliamentary representatives. The reduction in the power of Parliament could, he continued, be more than compensated for by the appointment of a select group of 'experts' to advise government. This, he believed, would greatly increase the efficiency of the nation and empire.

Berry's alarmism concerning the threat to national efficiency posed by the mentally deficient struck a chord in the community. For example, upon the announcement of his departure in 1929, the influential Melbourne *Herald* in an article titled 'MENACE OF THE MENTALLY DEFECTIVE: PROPOSAL FOR CENTRAL CHILDREN'S COURT AND TESTING CLINIC: DANGER OF LOSING PROFESSOR BERRY', demanded that Parliament act promptly to set up the clinic requested by Berry to investigate and treat the mentally deficient, otherwise he would leave Melbourne. It continued: 'The *Herald* is able to state that after a trip to England and America, he will be prepared to continue his research work in Melbourne if a department is formed to give him the necessary scope'.[29] The *Herald* claimed that the need for Berry to continue his work was imperative as it:

> may be said that all doctors agree that the problem of the feeble-minded is the most important and vital social question of the present day. One medical writer has gone so far as to say that unless the birth rate of mental defectives is restricted the British race is doomed.[30]

Berry had powerful supporters in his efforts. Medical graduates were among his most enthusiastic followers, as was the *Medical Journal of Australia*. Four Melbourne graduates, Stanley Argyle (Minister for Health, and later Premier of Victoria and knighted); Jean Greig (Chief Health Officer for the Education Department and Royal Commissioner); John Cumpston (first Director of the Commonwealth Department of Health); and Harvey Sutton (eventually Professor of Population Health at the University of Sydney) were prominent among them. In 1929 Argyle told the Victorian Legislative Assembly when speaking of the reintroduction of a Bill to deal with mental deficiency drafted by Berry, that 'Professor Berry's work is recognised the world over' and that 'the work he [Berry] did in this matter will live after he has gone'.[31] In 1919 on one of the many occasions it supported Berry, the *Medical Journal of Australia* proposed that strong measures be taken to eradicate the mentally defective population, including homosexuals and prostitutes: '[w]e have pleaded for a proper reform of this nature. We have insisted on the fact that this problem is a medical one and that it would be doomed to failure if entrusted to laymen'.[32]

Almost without exception, the Melbourne press supported Berry's views. In July 1923 he was interviewed by the *Sun News Pictorial*, in which he outlined his plans for segregated institutions for the mentally deficient:

> "You see," he adds worriedly, "all the time that nothing is being done to stop them they are increasing! They breed like rabbits – more of their own kind!" "What do you think," he is asked, "of the suggested medical certificate before marriage?" He shakes his head. "No", he says, "people won't have their marriages interfered with! Not yet, at any rate! Still, they'll come to it."[33]

The *Age* (the premier progressive Melbourne daily) was one of his strongest supporters. On 3 August 1928 the editorial stated:

> Persons having some semblance to rational human beings, but incapable of assuming life's ordinary obligations, are in existing society at perfect liberty to marry and to furnish a fresh brood, usually of a worse type, to be a financial burden and a social eyesore to their healthy-minded contemporaries ... [the] physically weak child is no problem; environment and nutrient may transform him. But that weakness which is mental as well as physical, and which demonstrably has its roots in heredity, is quite another category. The abnormal fecundity of the feeble and unfit is a thoroughly established fact [and the] ... wastrel, the criminal, the epileptic, the inebriate can in instances innumerable be explained in terms of the formula "mentally defective".[34]

The *Age* believed that until then the Australian public had looked on the problem of the mentally defective 'impotently and seemingly contentedly'. It now sensed a change in this position as it believed that 'society, however, is becoming more articulate, more emphatic, in challenging their right to procreate'.[35] Many articles to this effect appeared regularly across the print media in the early decades of the twentieth century.

Berry harnessed the popularity of his views for the purpose of raising money for research; a task in which he was very successful. The Edward Wilson Trust of the *Argus* newspaper provided him with significant funds in order to support a laboratory based in the Children's Hospital, and specifically in the Anatomy Department. In December 1925, a three-year grant of £300 per annum was made available for Berry to continue his research with further grants of £300 in 1927 and £850 in 1928, both for the purpose of hiring additional staff.[36] To understand the significance of this funding, in 1925 only £280 was allocated for all other medical research in the entire University.[37]

National efficiency (sometimes known as social efficiency) was his catch cry, as it was for the growing middle classes, and anatomy was to be the tool he would use to convince Australian governments to implement public policies designed to improve the race, and thus increase national efficiency. How could this hidden, feeble-minded or mentally deficient class be recognised? What was its size, and why did it constitute such a major threat to national efficiency and racial vigour? And further, what was the role of anatomy in the cure?

Who were the mentally defective?

In the opinion of contemporary academics there were four different grades of mental deficiency: idiots, imbeciles, the feeble-minded and moral defectives. Idiots and imbeciles were defined as not being able to care for themselves and their treatment was taken as coming under the *Lunacy Act*.[38] The Tasmanian psychologist and literary critic (later Vice-Chancellor of the University of Tasmania), Edmund Morris Miller, provided the standard legal definition of 'higher' levels of mental deficiency (the feeble-minded) in a paper presented to the 1924 meeting of the Australasian Association for the Advancement of Science:

> Persons in whose case there exists from birth, or from an early age mental defectiveness not amounting to imbecility, yet so pronounced that they require care, supervision and control for their own protection or for protection of others or in the case of children, that they by reason of such defectiveness appear to be permanently incapable of receiving proper benefit from the instruction in ordinary schools.[39]

This was a definition that would be repeated in numerous forums. Argyle used the definition in Parliament and, when explaining 'what science is actually teaching today',[40] pointed out that feeble-mindedness was generally agreed upon by '(s)cientists, educationists and the medical profession' to result from 'an arrest of intellectual growth in the early developmental life'.[41] The causes, according to Argyle, were primarily due to inheritance but could also be environmental deficiencies such as prenatal neglect, disease or injury. In the Victorian Parliament in 1926 Argyle warned that if a degenerate individual 'reproduces its own species, that deficiency will be transmitted. One can develop downwards not upwards'.[42]

According to Berry, mental deficiency was primarily associated with small head size. As a consequence, anatomy was the most important tool in recognising the pathology. In his memoirs Berry wrote that throughout his career he had felt:

> called upon to tell the story of what I am sure the good Dr Watson would have called the League of the Small-headed Men. (The League has been with us since the beginning of time and I fancy that Cain must have been one of the foundation members.) As constituted today the League consists mostly of unimportant people, whose activities are attended by no benefit to others. In fact the League is responsible for most of the crime of the world.[43]

Berry argued that mental deficients were distinguished from the insane as deficiency (amentia) and insanity (dementia) were generally seen to be distinct phenomena. Berry and his followers, however, believed that the small brains of the feeble-minded were less likely than average size brains to withstand shock, thus being disproportionately represented in asylums and among those soldiers suffering from shell shock after the Great War.[44]

How many?

The estimates of the numbers of mental defectives in the community varied but the incidence was generally felt by experts to indicate a significant problem. Most research had also claimed to establish that the incidence of mental deficiency was similar in Australia to that in other developed nations and, as Australia had one of the best living standards in the world, environmental considerations seemed to be insignificant.[45] Berry's friend and colleague, the Inspector General for the Insane in Victoria, William Ernest Jones, conducted a nationwide survey for the Federal Government in 1928–9. The nation's teachers categorised virtually every schoolchild in the nation as being either normal or mentally deficient. Jones claimed in his final report that 3.2 per cent of schoolchildren in Victoria were mentally deficient.[46] Berry's research claimed that a very large proportion of criminals could be classified as such.[47] And he privately stated that ten per cent of the normal population was feeble-minded.[48] The *Argus*, writing of Berry's report in 1929, claimed:

> [i]f by mental deficiency is understood only obvious idiocy and imbecility, its incidence might be covered by ½ a per cent. Mental deficiency, however, is much more widely spread. It includes all those cases, many of which are unrecognisable under ordinary medical examination, which have a degree of cerebral under-development, the consequences of which, to the individual and to society, often depend on environment.[49]

This alarmist view was repeated in the medical press. The leading article of the *Medical Journal of Australia* in November 1931 stated:

> [t]here is no need in a journal such as this to refer to the great army of mental defectives in the community; to their association with habitual drunkenness, prostitution, venereal disease and crime; to their prolificacy; nor to the economic aspect.[50]

In 1934 Berry claimed, in Britain, that for sterilization to be effective in eliminating mental deficiency 'one quarter of the population would have to be sterilised'.[51] As mental deficients were perceived to be breeding faster than the normal population, the problem was represented as one of national importance. Berry had previously argued in the *Argus* in July 1923 that the mentally deficient 'reproduced their kind from two to six times more than normal people'.[52] As late as 1939, when comparing the incidence of mental deficiency in the community in 1926, Argyle claimed that he thought the problem of mental deficiency 'was acute then; it is probably still more acute today'.[53]

Background to Berry's research

Though totally discredited later, Berry's research was the result of a combination of two related and very influential nineteenth century academic traditions, both traceable to Charles Darwin and his circle, and both intimately connected to the study of anatomy. The first influence was that of the new academic discipline of Physical Anthropology, especially one of its central tenets that race and class differences were primarily expressed biologically, not culturally. Head shapes, physical form and colour were to become the only certain ways of measuring the racial origin of any individual or classifying a race. The period covering the second half of the nineteenth century and the beginning of the twentieth century has been called the age of racial measurement and at its core was Physical Anthropology. By 1899, the American sociologist, William Z Ripley, claimed that over one and a half million adults and ten million children had been measured in Europe and the United States for their racial identity.[54] Physical anthropology also became institutionalised in many universities. There were numerous public policy consequences of the rise of this new discipline. For example, it was at this time that the races of Europe were categorised into fixed groups—Teutonic (Aryan), Alpine and Mediterranean (eastern, central and southern). These definitions became important mechanisms of choice in Australia's immigration policy, especially after the Great War, with a strong bias against Mediterranean immigrants. And some even went as far as to posit separate biological origins for the middle class and the 'degenerate' slum dwellers of the cities. The English writer, Henry Mayhew, in his study of the London poor, *London Labour and the London Poor* (1861–2) wrote of the poor as wandering tribes or races, able to be distinguished from the civilised both physically and behaviourally.

The notion that inferior races and inferior individuals within races had a smaller brain size started with the American anthropologist, Samuel Morton, who filled skulls with seed or shot to measure their capacity—one of the methods used in Melbourne by Berry and his research assistant, Stanley Porteus, when measuring unearthed Aboriginal skulls.[55] With heads they collected from unclaimed bodies in the care of the City Coroner and the Melbourne Hospital, other methods were utilised to correlate brain size with skull size. Porteus described these methods:

> since time was indeed of the essence of the matter, we both became very adept at the post-mortem method of removing the brain by sawing through the skull along the standard line of dissection of the scalp, lifting up the skull cap and cutting the brain stem. Suspended by a string, the brain was then carefully lowered into a full glass container and the overflow measured. Over a period of almost two years, we accumulated a series of over thirty brains.[56]

In Australia, Berry argued, it was necessary not only to acquire the best races but also the best examples of the best races to populate the empty continent, and this was his rationale for measuring skulls. Berry was an enthusiastic researcher in the discipline of Physical Anthropology, and claimed it was his discipline that could provide the selection criteria to ensure the best possible population would fill the empty continent. In 1924, he wrote in the *Herald*:

> What gardener develops his weeds at the expense of his plants? What cattle-breeder rears his animals from degenerate stock? Surely a human soul is

> worth more than a weed or a mongrel. Where, then, is the politician with the necessary vision to see that his bolstering up of the unfit at the expense of the fit is the worst form of national gambling, inasmuch as it means the State is perpetually backing the wrong horse? [57]

As an important initiative he advocated the establishment of a Chair of Physical Anthropology in Melbourne, as he told a Ballarat audience in 1912, so that:

> Eugenics, Mendelism or scientific breeding, abuse of charities ... and other problems of such vital interest to the well-being of our people could be studied and investigated.[58]

A discussion of physical anthropology and the collection of Aboriginal remains will follow in Chapter Nine.

The other intellectual influence on Berry was the anthropometry and eugenics of Francis Galton and his pupil, Karl Pearson. Berry was a great admirer of Pearson's statistical methodology, and had been invited to work with him at the Galton Eugenics Laboratory at the University of London, where Pearson was the first professor.[59] During the First World War, Berry had also been offered the Chair of Anatomy in King's College at the same University, which would have placed him in Pearson's intellectual milieu, including leading intellectuals such as those of the Fabian Society.[60] Berry's association with anthropometrical research work in Britain was such that he was invited to celebrate with colleagues the 1927 anniversary of the founding of University College, London, and to attend the celebratory dinner with Pearson at the Galton Laboratory.[61] Berry's research in Melbourne was an attempt to verify some unfinished earlier work of Pearson's on the correlation between head size and intelligence. Pearson's name is still well-known and acknowledged in statistics today, as he established the use of the correlation of numerous physical attributes to create a population profile producing the so-called 'bell' or normal curve. In 1903 he attempted to prove the genetic basis of intelligence in experiments by establishing a correlation between various physical attributes that presumably were genetically acquired (such as eye and hair colour), and the categorisation of the students' academic performance by their teachers.[62] Pearson abandoned this project because

The eugenics laboratory

Throughout the first three decades of the twentieth century Berry conducted his research campaign at a laboratory at the Children's Hospital, the psychological laboratory at the Melbourne Teachers' College and the Anatomy Department in the Medical School. Diagnosis of mental deficiency involved three groups of tests, apart from craniometry. First there were physical tests which involved an estimation of the cubic capacity of the brain and the measurement of the subject's weight, and standing and sitting height. Berry claimed that seventy-five per cent of the feeble-minded (or 'aments' as he called them), failed to attain the normal physical average for their age.[69] The second group was called psycho-physical tests and according to Berry 'although all three were primarily physical tests ... they were to a certain extent controlled by the nervous system'.[70] These involved measuring the handgrip and vital capacity of the lungs (the amount of air that could be inhaled) and, as only eight per cent of the mentally defective reached the normal average, this measurement 'could safely be taken as a significant, reliable, single index of mental capacity'.[71] Finally, there were the recently devised Binet-Simon written IQ test and Porteus' newly devised maze test, which Berry used in conjunction with the others, though he was dubious about their worth.[72] From these results, Berry made a professional decision as to whether the patient was feeble-minded or morally deficient.

The Director of Education, Frank Tate, supported Berry's research in numerous ways, but importantly he allowed Berry to measure the heads of ten thousand state school students. In 1918 Berry was pleased to announce that not only the Education Department, but also the headmaster of a private school in Melbourne, had 'put our advice to the practical test. He now systematically causes every boy's cubic capacity to be estimated'.[73] On the basis of his friendship with L A Adamson, it may well have been Wesley College. After receiving Berry's report, the unnamed private school sent a boy to the university for further testing. Berry's case study of this individual indicates the type of information he believed could be extracted from a child using his scientific method. His calculations showed that the student had a brain size norm referenced to that of a two-year-old and had 'no interest in schoolwork, and has made little progress in arithmetic and reading'.[74] What is more, he lied, stole,

Professor Richard Berry in front of the old Medical School.
Courtesy: The University of Melbourne Archives.

was a sex offender and 'a prey to every impulse'.[75] Along with an above normal psycho-physical average, an above normal physical average and a below normal Binet-Simon and Porteus test average, according to Berry, he was '(a) moron of low grade. Deficiency due to cerebral under-development ... morally irresponsible'.[76]

The public campaign

Berry tirelessly conducted an extensive public education and propaganda campaign that took him around Victoria, to each capital city in Australia and overseas. He claimed 'I suppose I have given more lectures and addresses on the subject of mental deficiency than almost any other living man.'[77] During his Australian campaign, he used the most up to date methods (including annotated lantern slides and a very early silent movie) to lecture in large halls filled with interested members of the public.[78] The Edward Wilson Trust of the *Argus* financially underwrote the cost of this tour. In 1926 Berry delivered the second of three Beattie-Smith lectures in the Medical School. In this widely reported lecture, Berry gave the example of the bushranger, Ned Kelly, as an archetypal moral and mental defective, claiming that his research had shown Kelly to have the brain size of a fourteen-year-old.[79] Another example given was the murderer, Frederick Deeming, who Berry claimed had the brain size of a thirteen-year-old.[80] Berry also argued that the measurements he had taken of the heads of criminals in Melbourne gaols had shown that their brains were distinctly smaller than average. He warned that 'the danger with this form of mental defect ... lay in the fact that it passed unnoticed, yet required very special diagnosis'.[81] As discussed in the first chapter, Halford examined Ned Kelly's skull phrenologically at the University of Melbourne. Clearly Melbourne's celebrity criminals, beginning with Harrison in 1864, were to have a long association with the Anatomy Department.

It was not just criminals that attracted Berry's interest. On Wednesday 4 July 1917, the Melbourne *Herald* informed its readers that on the following day a lecture entitled 'Brain Growth and Education', was to be delivered by Professor Richard Berry under the auspices of the Council of Public Education. The *Herald* believed that this 'lecture is of more than ordinary importance, inasmuch as it furnishes, for the first time in the history of human knowledge, information as to the capacity of the brain'[82] of the normal range for boys aged between six and twenty-one years. The lecture, according to the *Herald*, also indicated how:

> it may be possible by co-operation between the neurologist, the psychologist, and the educationist to pick out those boys who depart so much from the normal that they may not improbably be regarded as the future inefficients or potential geniuses of adult life and be treated accordingly.[83]

The article continued with an outline of the research to be presented by Berry at the lecture, pointing out that the study was based on the head measurements of 1,139 normal boys aged between six and twenty-one, drawn from state and public schools and the University. As the controls, Berry examined three hundred and fifty-five criminals, fifty-three deaf and dumb boys, thirty-nine 'mentally deficient' boys and sixty-four 'living Australian Aboriginals'.[84] In Berry's unpublished autobiography, he recalled giving this lecture:

> The researches of Mr Porteus and myself, dealing as they did with the school and the University population of Melbourne, naturally aroused a good deal of interest both among the people and the Press. In fact the first lecture we gave in one of the larger Collins Street halls attracted even the State Governor and his presence meant, of course, a full house.[85]

Under the heading 'LARGE AUDIENCE APPLAUDS LECTURE BY PROF. BERRY—BRAIN GROWTH AS SUBJECT—HERALD GIVES £150', the *Herald* reported on 6 July that the lecture held the previous night in the Presbyterian Assembly Hall had been delivered to 'a packed and attentive audience. Every seat was occupied, and many people stood in the passages the whole time'.[86] As well as the Governor (Sir Arthur Stanley), the lecture was attended by some of the most significant figures in State education in Victoria. The Minister of Education, H S W Lawson presided, and in the audience were George Swinburne (described as 'an Interstate Commissioner and a leading technical educationist'), Theodore Fink (Chairman of the Royal Commission on Technical Education 1899–1901 that had recommended the introduction of secondary education for all) and Frank Tate, Victoria's first Director of Education.[87] At the conclusion of the lecture, Fink moved the vote of thanks 'and remarked that the Minister for Education need not dig very deep down to find the solid substratum of people in favour of liberal education expansion'.[88] Swinburne, in seconding a vote of thanks to Berry, complained that it had been:

> stated that only one child in 20 was capable of education after the age of 14. This was a view which he did not think any democratic community would accept, and Professor Berry's lecture had proved that such a view was wrong.[89]

As well as the précis of the lecture in the *Herald*, a complete version was published in the *Medical Journal of Australia* following a presentation of Berry's and Porteus' research findings at a meeting of the British Medical Association on 2 May in Berry's Anatomy Department at the University.[90]

Berry advanced two main conclusions in his lecture. The first was that while head size alone could not be used as an indication of intelligence, among a large group of individuals there was 'a distinctly measurable correlation between size of head and intelligence'[91] and while he 'did not deny the possibility of brain quality entering into the problem ... he could find no proof of it. The facts seem to point to an association between size of brain and mentality'.[92] The second conclusion reached by Berry was that in any comparison between different classes in society 'there is a distinctly measurable correlation between size of head and intelligence'.[93] The order of groups in Berry's research population (in descending order of head capacity and intelligence) were:

- Educated class groups, for example, British university teachers, university students and the average member attending a meeting of the British Association for the Advancement of Science;
- Schoolboys and university students, aged from six to twenty-one (private schoolboys had larger heads than state schoolboys);
- Socially inefficient groups such as criminals;
- Abnormal classes such as the mentally deficient and the deaf and dumb, and
- One primitive and non-caucasian race – the living pure-bred Australian Aboriginal.[94]

One of Berry's lantern slides of a Melbourne criminal 'Gyp the Blood'.
Courtesy: Royal Australasian College of Surgeons Archives.

As can be seen from the above groups, Berry was advocating a radical reordering of social merit based on intelligence as expressed by head size. He had noted shortly after arriving in Australia in 1905 that the rigid class system in England had been weakened in the colonies. As previously discussed, Berry believed that the potential for creating an aristocracy of intellect was greater in Australia than Britain, due to this dilution of the English class system.[95] Victoria and Australia therefore offered an opportunity for Berry's new classification of intellectual merit to be instituted into the class structure. He concluded that eventually the medical scientist would be able to categorise every person in society into these new classes based on merit. When this hierarchy of intelligence and head size was linked with Berry's assertion that 'it was usually accepted that no new neurons were formed after birth',[96] then education could only be for the purpose of sorting people according to their inherited biological ability.

Berry was of the view that the inferior classes did not obey 'the laws of Man but the dictates of Nature' and the evidence of the small heads of the lower of his classes exploded 'the common error of supposing that Education could succeed where Nature had failed, and that is, put brains where they are not'.[97] There was no point, according to Berry, in offering a wide range of opportunities to all students as the hierarchy of intelligence was established at birth, and became obvious by, at the latest, the twelfth year. Berry's conclusion supported a graduated and multistreamed system of secondary education providing education for each citizen to the age of sixteen, at which stage, according to Berry, the brain would constitute eighty-two per cent of its final adult size. Brighter children (classified as members of Berry's superior class), would receive an education from the age of twelve that would lead to university and the professions. The lower classes would receive a technical education that would include training to be a good citizen. The mentally deficient (which, according to Berry, included many of the criminal class), should be institutionalised for the purpose of restricting their right to breed, and thus preventing them from perpetuating their faulty genetic make-up. Berry concluded his lecture by rhetorically asking whether there was an urgent need to establish a laboratory in order to coordinate:

> the services of the anthropologist, the psychologist, the educationalist and the medical man, in a great national endeavour to discover from among the school inefficients the future social inefficients of adult life, and to do so in the only possible way, by the foundation and maintenance of a special laboratory for that purpose?[98]

Following Berry to the stage, Porteus presented evidence in support of Berry's research findings and told the audience:

Berry's office. Courtesy: Medical History Museum, The University of Melbourne.

> They [society] were in the habit of applying the same system of education to all children, regardless of their different intellectual endowment. The future criminal and the potential genius might be found side by side in the same class under the same teacher.[99]

Berry's success

As the states rather than the Commonwealth controlled health under the Constitution, the best chance Berry had for implementing his ideas through the medium of public policy, was the three Mental Deficiency Bills presented to the Victorian Parliament in the interwar years.[100] Similar Bills were introduced into other State legislatures. The first and only one enacted was the *Tasmanian Mental Defectives Act 1923*, drafted by Berry.[101] The Tasmanian Act was relatively ineffective because there was not the financial underpinning needed to fund the widespread institutionalisation recommended. In essence Berry's Bill was followed by all the states, and was based on the United Kingdom's *Mental Deficiency*

Act 1913. Its aim was to institutionalise the mentally deficient in the population rather than sterilise them; sterilisation being the procedure followed in the United States and Scandinavian countries, where up to sixty and ninety thousand citizens were correspondingly 'treated'.

Berry's Bill was plagued with bad luck in Victoria. It was introduced three times—in 1926, 1929 and again in 1939—after the outbreak of the Second World War. In each case, the Bill was passed unanimously by the Legislative Assembly (the lower house) and in the case of the 1939 Bill, both the Legislative Assembly and the Legislative Council (the upper house), though the legislation was never proclaimed.[102] On the first two occasions the Parliament was dissolved due to unrelated gerrymander issues and the third Bill, though enacted, seemed to get caught up in the perennial problem of the Dunstan government's notorious inactivity and also because after the war, eugenics fell dramatically out of fashion.[103]

The Bill aimed to institutionalise a significant proportion of the population to prevent them from breeding. Though encountering no opposition in Parliament to any of these attempts to introduce the Bill, there simply was not enough money to build all the institutions needed in Victoria. Also, health spending plummeted during the Depression. Two institutions were set up in the 1930s at Travancore and Janefield, following the earlier establishment of two schools for mentally deficient children in Montague Street, South Melbourne and Bell Street, Fitzroy. They provided only a fraction of the capacity envisaged by Berry and his supporters.

Did Berry fail? In his old age he certainly felt he had. It must be remembered, however, that many characteristics of Victorian and Australian society in the interwar years conformed closely with his ideas. For example, Berry's campaign to stop the introduction of free university education was entirely successful, and the White Australia Policy was underpinned by the scientific ideas he preached.[104]

It also seems likely that Berry's research influenced his friend and colleague in reform, Frank Tate. The early decades of the introduction of secondary education into Victoria following the *Education Act 1912* are distinguished by the construction of Technical Schools in working class areas of Melbourne, from which no child could graduate to the University.[105]

Although rarely criticised, Berry's ideas did occasionally attract

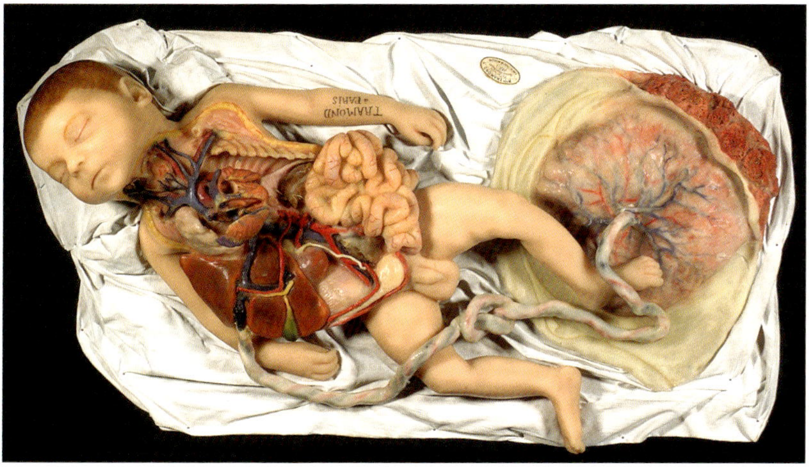

Baby with organs modelled in wax by Tramond, Paris, nineteenth century.
Courtesy: Harry Brookes Allen Museum, The University of Melbourne.
Photograph courtesy: Stuart Thyer and Michelle Gough.

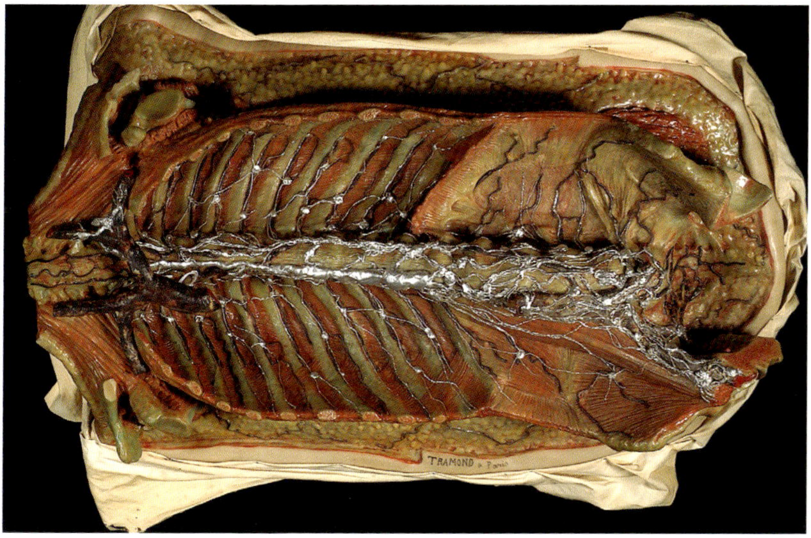

Lymphatic system modelled in wax by Tramond, Paris, nineteenth century.
Courtesy: Harry Brookes Allen Museum, The University of Melbourne.
Photograph courtesy: Stuart Thyer and Michelle Gough.

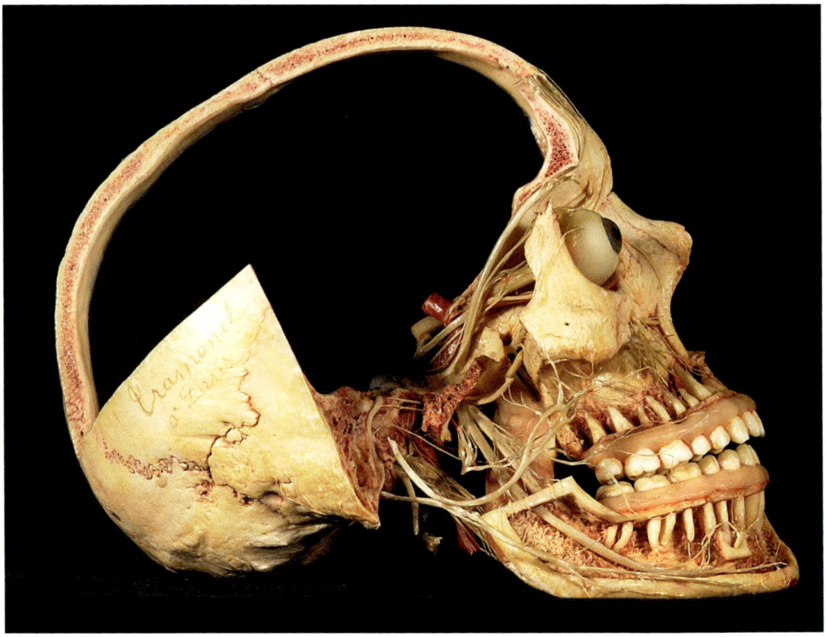

Skull modelled in wax by Tramond, Paris, nineteenth century.
Courtesy: Harry Brookes Allen Museum, The University of Melbourne.
Photograph courtesy: Stuart Thyer and Michelle Gough.

Painting of snake by Wood Jones when he was a medical student.
Photograph courtesy: Barry Christophers.

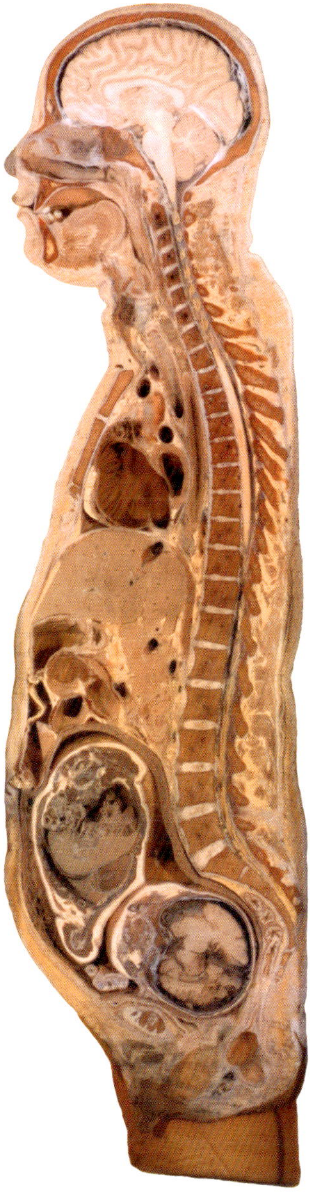

Sagittal section of pregnant woman, prepared for Richard Berry by William Preston.
Courtesy: Harry Brookes Allen Museum, The University of Melbourne.
Photograph courtesy: Stuart Thyer and Michelle Gough.

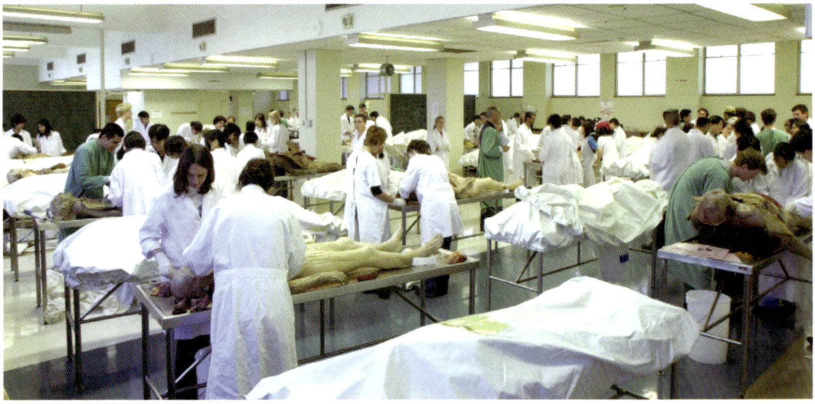

Dissecting Theatre in 2002. Photograph courtesy: Stuart Thyer.

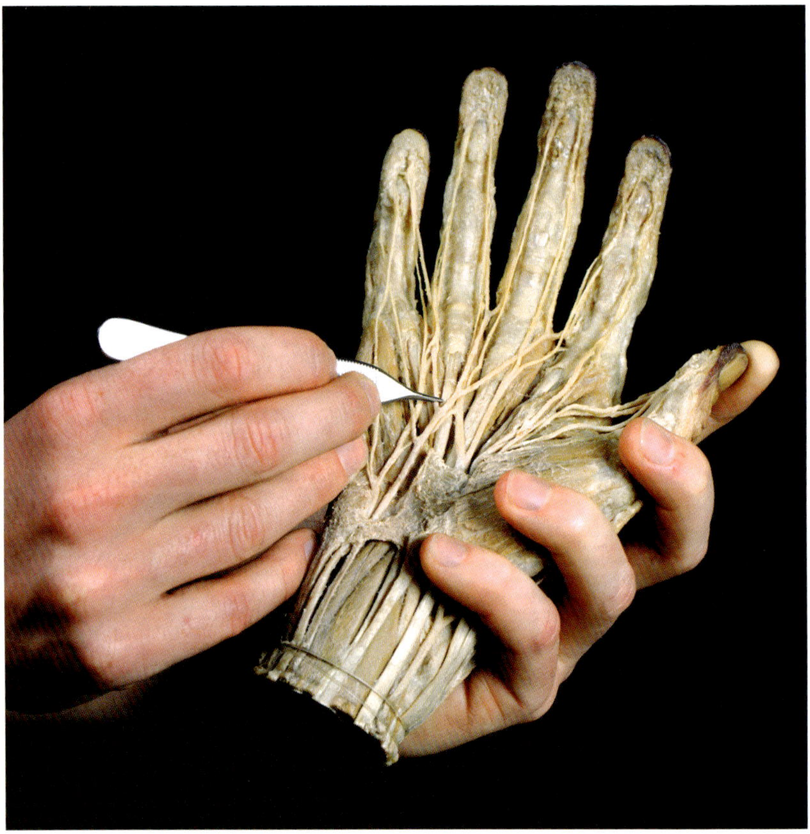

'Three Hands', twenty-first century dissection. Photograph courtesy: Stuart Thyer.

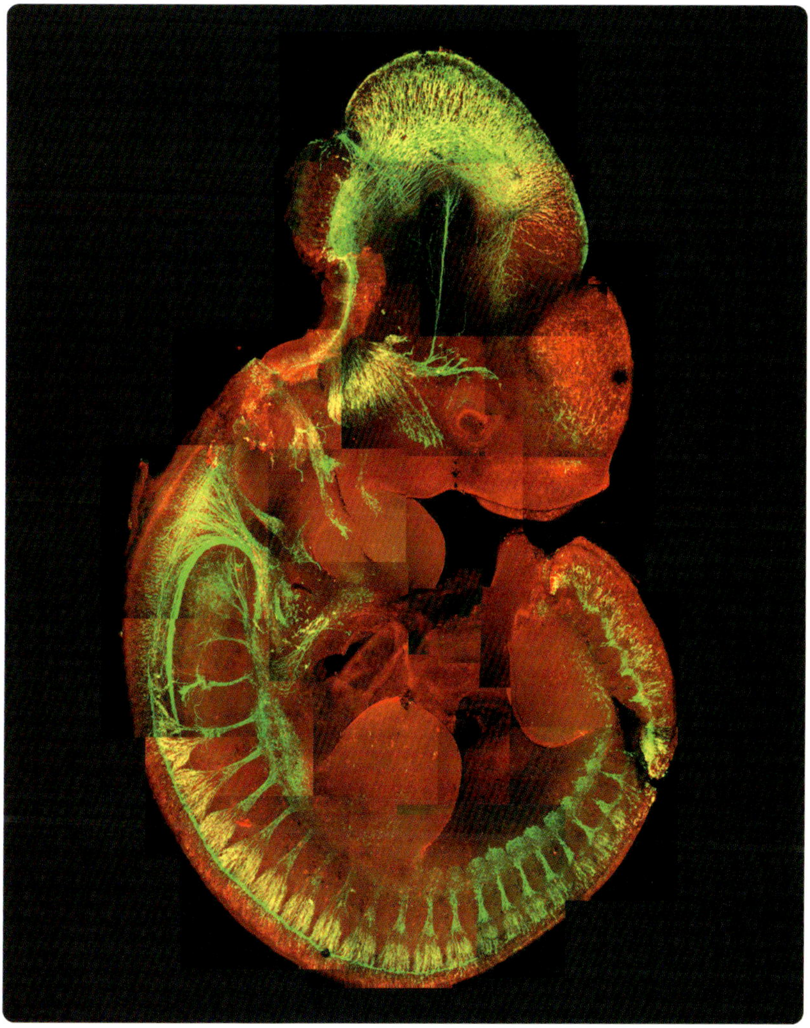

Montage of confocal microscope images of a whole mouse embryo, stained in green to show the developing nervous system. The developing brain can be seen at the top, and the developing nerves and ganglia throughout the body. The whole embryo is approximately three millimetres from top to bottom. Image courtesy: Ashley Stewart.

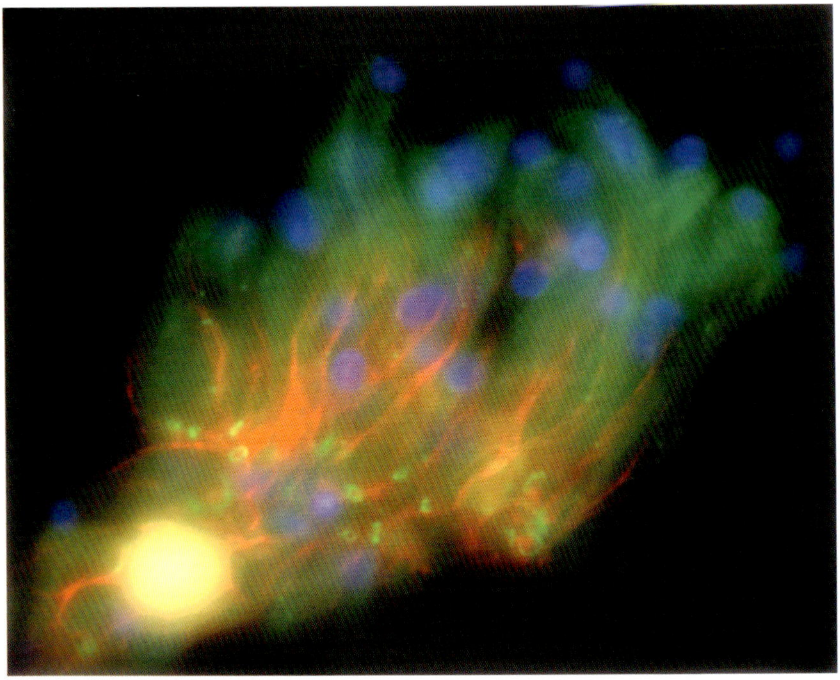

A developing cluster of sperm cells. The sperm heads are shown in blue and the tails in red.
Bright green rings indicate connections between the sperm cells.
Image courtesy: Gary Hime.

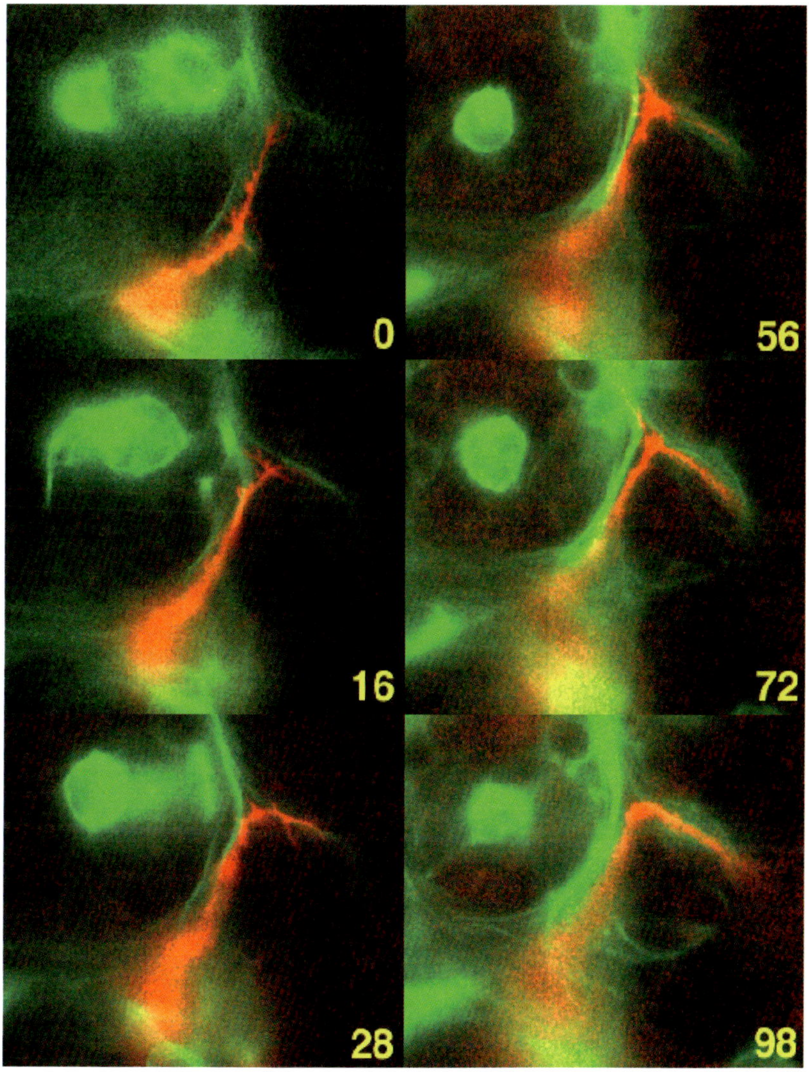

A sequence of images (taken over 98 minutes) of a neuron growing in a living embryo. The neuron (red) extends an axon past other cells (green). Eventually the axon turns to follow a small nerve branch to the right. The total length of the axon is about 40 micrometres. Image courtesy: Paul Whitington and The Journal of Neurobiology (1998) 37: 607-21.

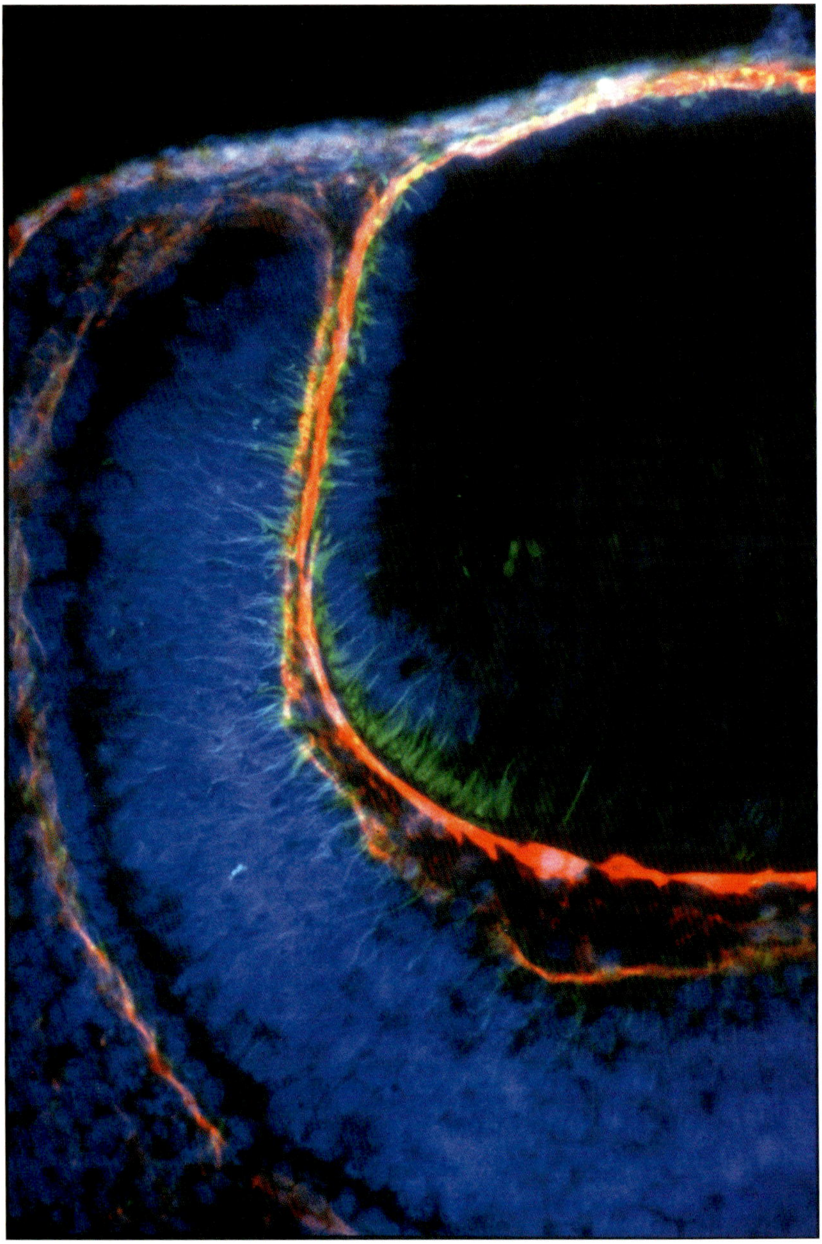

A section through the developing eye of a mouse embryo as seen on a fluorescence microscope. The large circle (top right) is the developing lens. It is cupped in the developing retina (to the left and below). The distribution of a key signalling molecule (FGF2) is shown in green, while the elastic capsule that surrounds the lens is shown in red. Image courtesy: Robb de Iongh.

controversy. In November 1912 the *Argus* reported on 'the interesting work … recently … carried out by Professor Berry'[106] that had involved the measurement of the cranial capacity of three hundred and fifty-five adult male criminals in the Pentridge and Melbourne gaols. In this research, Berry claimed that criminals had smaller skulls than well-educated members of the community (with anatomists topping the scale). Furthermore, he stated there was a similar hierarchy shown by the skulls of criminals according to the intelligence (head size) of the perpetrator from skilled crimes like forgery and embezzlement to cattle stealing.[107] The following week, the *Argus* took a substantial shot at Berry's findings in its editorial, lampooning the efforts of nineteenth century sociologists under the leadership of Herbert Spencer, to collect statistics of no obvious practical value:

> An army of serious students went out into the world to acquire sociological data, and came home with their bags full of statistics. These statistics they subsequently embodied in papers, read before learned societies, showing with exactitude the prevalence of freckles amongst bi-metalists, and reporting a decrease of left-handedness amongst dissenters. Members of the outside public were a little mystified by all this.[108]

The *Argus* went on to claim that the sociologist had been supplanted as an icon among the fashionable by the anthropometrist who was:

> a statistician armed with a measuring-tape and an imposing array of technical terms – microcephalous, macrocephalous, dolichocephalis [sic], prognathous, and a thousand more – by which the public is greatly impressed … Under the leadership of Professor Karl Pearson, the anthropometrists go up and down in the world measuring people, and especially measuring people's skulls. They have already accumulated a great deal of truth … if one wishes to know the average length of the eyelashes of unmarried female Eskimos.[109]

Having explained to their readers that Berry's research claimed to show that there was an appreciable correlation between head size and intelligence, the *Argus* then remarked that it was a very 'curious and pleasing' coincidence that two anatomists (Berry and his associate,

L W G Büchner) should have discovered that 'anatomists have the highest brain capacity of all the classes!'[110] The *Argus* attacked the assumption that anatomists had the most intelligence in all the groups measured and that inmates of the prisons represented the most unintelligent class in the community. On the contrary, it continued, 'any forms of crime imply intelligence far above the average'.[111] The *Argus* claimed that the child of respectable parents brought up among criminals in squalor would become a criminal, and vice versa, and that '[a] man's criminal act may prove him to possess mental energies which, rightly directed, might have made him a great soldier, a great statesman, or even [in an obvious dig at Berry] an anatomist'.[112]

As we have seen in earlier chapters, this was not the view of social Darwinist circles in Australia, and it was certainly not the view that held sway until at least 1939 in middle-class reform circles. With his usual form, Berry struck back immediately on the Monday following the Saturday editorial in the *Argus*. He observed that 'to merit the honour of a leading, or misleading, article in Melbourne's leading daily is fame indeed'.[113] He observed (in a comment that explicitly marks him as a social Darwinist), that if it had been only him who had been castigated by the lead writer:

> I should have felt myself duly chastened, but as I find that this same universal authority also corrects the scientific absurdities of Herbert Spencer, Cesare Lombroso and Karl Pearson, I feel that I err in the company of the immortals – whom it is obvious your "printer's devil" will never succeed in joining.[114]

In what was either a placatory or niggling note following Berry's letter, the *Argus* asked whether Berry would give a reference to a standard work on brain measurements as a test of sense of humour as '[w]e never realised until we read his letter how great is the need for some trustworthy authority on the subject'.[115] Whether the *Argus* meant to infer that the original attack on eugenics was merely a joke at Berry's expense is unclear. What is not surprising is that it was the *Argus* (the most conservative newspaper in Melbourne) that launched the only public attack on his ideas. Eugenics was progressive, reforming middle-class public policy.

Berry's departure

While Berry's views were widely listened to, and although he was a popular and highly effective teacher, he had made some significant enemies in the University administration and the medical profession. So much so, that in 1928 Berry was informed that the usual five-year extension of a professorial appointment to the age of sixty-five would not be followed in his case. This was almost certainly engineered by Barrett. Berry was very bitter about this, but he managed to obtain a well-paid post as chief medical officer of a large institution for the mentally deficient at Stoke Park Colony in Bristol, England; a place where he could continue his research. His final lecture in Melbourne was a riotous farewell party organised by his students who sang University songs with Berry. They then accompanied him to the port.[116] They undoubtedly would have performed 'The Anatomist's Song' with words by Berry and sung to the tune 'A more humane Mikado' from the operetta *The Mikado* by Gilbert and Sullivan. Like Allen before and Wood Jones after, Berry indulged in writing poetry:

> A more humane anatomist did never before exist,
> To nobody second, I'm certainly reckoned, a modern materialist.
> ....
>
> My object all sublime, I shall achieve in time –
> To make anatomy up to date, anatomy up to date!
> I'll make each student gaze,
> In twentieth century ways
> On the bones of his lecturer's skeleton
> By means of the Roentgen rays.[117]

Berry certainly was a materialist, and this was pointed out in 1928 by the reviewer of Berry's *Brain and Mind*: '*No neurons no mind* is Professor Berry's favourite slogan.'[118] His materialism and eugenics were to be strongly attacked by his successor, Frederic Wood Jones.

Berry had entered into a University pension scheme in 1924, and to his great delight, after retiring at just beyond sixty, he continued to receive payments until his death at the great age of ninety-five, thus Berry had taken his revenge on the Council.[119] He also continued to have considerable influence in medical circles and beyond. He was made one of two colonial representatives on the British Medical Association, and shortly after he chaired a subcommittee that produced a report on the problem of mental deficiency in England. In 1932 the group reported that this problem was serious and growing.[120] Lionel Penrose, the eminent English geneticist and psychiatrist, described how Berry (the 'eminent anatomist') was the leader of an influential school of investigators in the 1930s that ascribed virtually all mental deficiency to 'rotten heredity'.[121]

An assessment of Berry's research raises interesting and pertinent issues. Eugenics, physical anthropology and, to a lesser degree, craniometry were fashionable topics in the early years of the twentieth century.[122] Berry's work received a disproportionate amount of research funds due to the perception that it was socially useful. Suffering the curse of living long enough to see a major reversal in fashion, Berry became embittered when he saw his work eclipsed and ignored, even despised and ridiculed. This about face is perhaps best illustrated in the University's student paper *Farrago*. In its very first edition, in 1925, Berry was front-page news. On this occasion he received attention because of his lecture delivered in the Anatomy School lecture theatre to the University Science Club on the subject of 'feeble-mindedness'.[123] His message, that the subnormal members of society were spreading their inferior genetic material, needed no hard selling at the Club at this time. The meeting was so well attended that there was standing room only. According to the reporter, the subject matter of the lecture awakened in the students:

> (t)he particular danger to the community of the moron – the type of individual who, although capable of managing his affairs in a more or less satisfactory way, is nevertheless definitely sub-normal and has not full control of the animal appetites and desires.[124]

Berry's slides were of the 'notorious' American Kallilak family[125] which the reporter believed would:

> provide the most terrible argument against present methods of assisting the feeble-minded to take their place as members of a normal community, instead of caring for them under conditions which will prevent them from sending into the world fresh supplies of sub-normal blood.[126]

In the second issue of *Farrago* in 2003 a call was made to rename the Berry building, due to the leading role he played in the eugenics movement.[127] Surely Berry's fate emphasises the vagaries of intellectual fashion, and it provides a cautionary tale to all who engage in medical research.[128]

CHAPTER SIX

Frederic Wood Jones, polymath: 'whoever wins to a great scientific truth will find a poet before him in the quest'

On 5 June 1931 Frederic Wood Jones, Professor of Anatomy at the University of Melbourne, delivered the Joseph Bancroft Memorial Lecture in Brisbane 'The Changing Point of View'. There must have been an ironical edge to this title, as Wood Jones spoke in the middle of a year in which the subtropical capital of Queensland experienced some of its worst ever floods. The scope of the lecture was wide, as was usually the case with Wood Jones at this time of his life, and included many historical musings and criticisms of trends in the contemporary world. He deplored 'the passing of that fine type, the medical man, who was, in every sense of the term, a liberally educated, deeply learned, yet generalised naturalist' having been replaced by 'the present day scientist ... a clean shaven, hard-faced, somewhat inhumane person, who, cloistered in a laboratory, holds a test tube to the light and sees in it Nature's profoundest secrets.'[1] He also pursued one of his lifelong obsessions—his intense dislike of the contemporary idolatry of Darwin—as, according to him, 'Darwinian evolution, from being a theory, became a creed.'[2] From 1930 to 1937 in Melbourne, and after nearly thirty years of research and publication in the disciplines of Anatomy, Ecology and Anthropology, Wood Jones created an alternative world view to that fashioned by Darwin's followers.

Although sometimes overly nostalgic for the past, he was no rigid conservative; rather he had become a deeply learned maverick by 1931. Apart from his unsurpassed reputation as an anatomist, he was also a Lamarckian when such a position amounted to scientific heresy. Wood Jones was also the author of a radical theory of humanity's evolution from creatures more primitive than the apes. He was an early conservationist, and a savage critic of many contemporary scientific racial theories.This had the effect of making him one of the rare supporters of the Australian Aborigine. Most of all, he hated the totalitarianism of intellectual orthodoxy. At the conclusion of his time in Melbourne, he wrote:

> In truth I think we are all sinking to the common level of a standardised loss of individuality. Few are prepared to let their fellows know that they are non-conformists, in a world that accepts commonplace mediocrity as a hallmark of respectability.[3]

Wood Jones's Melbourne sojourn—leaving aside six months leave as temporary Director of Anatomy for the Rockefeller Foundation at Peiping Union Medical College in China, from the last term of 1931 to the end of first term 1932—was significant. Not only was the duration of his stay in a largely peripatetic life unusual, but additionally it represented his major period of intellectual consolidation and maturation. The years that preceded Melbourne (from his beginnings as an academic adventurer in the Cocos Keeling Islands in 1904 to those as a Fellow of the Royal Society, leaving Hawaii for Melbourne in 1929) had seen him produce works of brilliant scholarship in the separate disciplines of Ecology, Anatomy and Anthropology. In Melbourne he began to unite all these strands into a unified theory of life, and these are best seen in his essays in *Life and Living* (1939). As such, it is impossible to understand the reason for his coming, the excitement he generated during his time in Melbourne, or the manner of his leaving without some understanding of his whole life's work. In the period following Melbourne until his death in September 1954, he was distinguished by more publications on his various interests.[4] Other essays from this last period attempted a unifying theory of life. And while developing themes that he had begun in Melbourne, these latter essays could be distinguished from his earlier Melbourne work because of his deep pessimism, brought on by the Second

World War.[5] Thus Melbourne can be seen as the central, unifying and perhaps most optimistic period in the latter part of his career.

Early life

Frederic Wood Jones was born on 23 January 1879 in West Hackney, Middlesex, the only son, along with two sisters, of Charles Henry Jones and Lucy Jones (née Allin). He described himself as 'an ordinary boy from an ordinary school in an ordinary suburban place.'[6] Indeed, he had trouble passing the matriculation examination for London University.[7] His father was a local slate merchant, builder and architect. 'Freddy', as his friends knew him, attended Enfield Grammar, and then studied at the London Hospital Medical College, later incorporated into London University in the year of his graduation (MB, BS) in 1904 (he had graduated with a BSc in 1903). It was while studying anatomy that he formed an intimate and lifelong friendship with the famous anatomist and anthropologist, Sir Arthur Keith, thirteen years his elder. As an undergraduate, and in contrast to his school days, Wood Jones blossomed into a brilliant student winning numerous prizes, and his work was published in academic journals.[8]

He was fortunate to be attracted to anatomy, as it was during his lifetime in the immediate post-Darwin period that the intense search for the origin of humanity had begun—anatomy being the principal tool used in this stage of the search. Exciting new discoveries of human remains resulted in heated battles over competing evolutionary theories. Wood Jones and Arthur Keith were active and important participants. To varying degrees they both picked the wrong horse and Keith's support of Piltdown man,[9] perhaps the greatest scientific hoax of the twentieth century, was the most damaging. For his part, Wood Jones erred (at least according to current thinking) by vigorously denying the close connection between Man and the great apes. He did, however, reject Piltdown man.

Adventures in the Indian Ocean

Upon finishing his degree in 1904 and with Keith's encouragement, Wood Jones relinquished the immediate possibility of an academic career and instead, he took the unusual step of becoming the resident medical officer for the Eastern Extension Telegraph Company on the Cocos Keeling Islands, remote atolls in the Indian Ocean. This he claimed, when questioned about the wisdom of such a career move, was for the sake of pure adventure. Thus began a life of wandering and although highly prone to seasickness, he was to spend over twelve months at sea.[10] It was during this time in the Cocos Keeling Islands that Wood Jones confirmed his love of the study of nature. He wrote to Mrs [Cecelia] Keith on his way to the Indian Ocean that 'really medicine is cramped and small when compared to ecology – pity one couldn't make a living of it.'[11]

Under Australian control, these remote U-shaped atolls truly tested his adventuresome spirit. Halfway situated between Australia and Sri Lanka in the northern Indian Ocean, the atolls are one thousand kilometres from Java and eight hundred kilometres to the nearest landmass (Christmas Island). There are only two habitable islands among the group, with a high point of only fifty metres and regular cyclonic storms. Of less concern to the ascetic Wood Jones, however, was that they provided only a meagre supply of food, discounting coconuts, rats, crabs and fish.[12] Wood Jones wrote to Arthur Keith in 1905:

> It's a never ceasing wonder to me how the deuce they [the inhabitants] stay here. If you watch the mountainous high rollers of the Indian Ocean swept on the S.E. wind you can't think why the Ocean doesn't get the land back again in the next 24 hours. When you consider that my house is on land raised 1 foot above the ocean level and the coral barrier alone resists the surf ... you feel that your grasp of the globe is not a strong one. And the sight of the ocean rising actually through the island where the island level is lower than the ocean level is apt to make you wonder if a 200 mile swim with sharks or a climb to the top of a coco palm is the better exercise.[13]

Whether by chance or design, this group of atolls was the only such geographical formation visited by Charles Darwin on his five-year voyage on the Beagle. He arrived at the atoll and spent ten days exploring in April 1836. It was here that Darwin formulated his theory about the formation of atolls, supporting the theory of his older mentor and later friend, Charles Lyell. Lyell had argued in *The Principles of Geology* (Darwin had taken a copy with him on the voyage) for a gradualist view of the changing geology of the world, that is, great changes could occur by adding together small changes over long periods of time. Darwin had collected fossil shells from fourteen thousand feet above sea level in the Andes, and speculated that this must have taken an enormous period of time to occur. Perhaps, he mused, the corals he saw at the Cocos Keeling Islands grew partly in response to the slowly changing geological landscape. Thus, he argued, coral reefs grew upwards on top of submerged, mostly ancient and inactive volcanoes as the ocean floor slowly subsided. This observation also supported Darwin's theory of evolution by confirming the concepts of deep time and a gradually evolving landscape, that is, there was enough time for evolution to occur. This was an important moment in the history of science. Huxley wrote 'I cannot but believe that Lyell was for others, as for me, the chief agent in smoothing the road for Darwin'.[14] And the poet, William Cowper, wrote:

> Some drill and bore
> The solid earth, and from the strata there
> Extract a register, by which we learn
> That he who made it and revealed its date
> To Moses, was mistaken in its age.[15]

Wood Jones spent fifteen months studying the reefs, examining the coral-building polyps and the base that supported them and measuring their rate of growth. From this research he concluded:

> There isn't the slightest doubt that Darwin's explanation of atolls and barrier reefs [formation] is wrong. This place is not and I doubt if ever has been, undergoing subsidence.[16]

And he did not find any evidence to change his mind. In fact, Darwin had been basically right about most atolls and coral islands, but wrong about the only one he had ever seen. Rather than being volcanic, Wood Jones had observed that the Cocos Keeling group of islands had formed on a sedimentary ridge. He stated:

> It is on the other hand quite easy to see that the whole thing is formed of wind and wave on an elevated ridge of ocean bed. A ridge by the way that extends probably to Christmas and certainly to North Keeling [islands].[17]

> I urge you to ... write an anatomy which will be something more than the mere record of dull morphology. I am especially anxious to have you do this ... I am hoping within the next three years to start to teach anatomy in a way that would make the older anatomists' hair rise.[28]

For Wood Jones, anatomy was inseparable from function. Even the application of formaldehyde in the late nineteenth century had proved problematic for him in this regard:

> the formalin-hardened cadaver was dissected with fresh zeal, and a new descriptive anatomy, in many ways the worst yet accomplished, was promulgated. Artificial and geometrical descriptions were given to viscera ... Surely here is the low water mark of anatomy as a science.[29]

Marriage and race

The Cocos Keeling Islands were to become central in the field of race science to Wood Jones, both personally and intellectually. At the time of his expedition to the Indian Ocean George Clunies Ross governed the atolls, since the British Government had granted his family governance early in the nineteenth century. Only a few hundred people inhabited them but many races were represented in the island's population.[30] The Clunies Ross family had married into aristocratic Malay families with the result that George was half Malay, and as he had followed his father and married into a similar family, his children were three-quarters Malay.[31] The race factor became significant for Wood Jones when he fell deeply in love with the third of George's five daughters, Gertrude, who had been educated in France and England. Unfortunately for Wood Jones, Gertrude was already married to Wilhelm Blom, a Scandinavian sailor employed by George, and mother to four of his children (a fifth child was born while Wood Jones was living on the island). The question of race in the British Empire at that time made Wood Jones's love for Gertrude unconventional at best. And the fact that she was already married made it even worse.[32]

Upon arrival on the Cocos Keeling Islands he had expressed in correspondence to Arthur Keith the standard view of race held by the Victorian middle classes:

> The natives ethnologically are no good. They are too mixed. Balinese … [and] various other Javanese races, Dyaks, Chinese and Turks – the remnants of a private gentleman's harem. All these races have mingled in the Cocos native.[33]

His later rejection of this contemporary attitude to race seems initially to have been the result of his growing love for Gertrude. This experience was to be the stimulus for his championing of Aboriginal rights in Melbourne, one of his major projects in the 1930s when he was Professor of Anatomy.

The year 1910 was especially significant for Wood Jones, for in addition to the publication of his first book and the awarding of his doctorate, on 11 June he married Gertrude in London at the Royal Chapel of the Savoy ('the Mecca of Bohemia'[34]). The occasion was the culmination of elaborate and expensive planning due to Gertrude's recent divorce. George Clunies Ross's settlement of a substantial sum of money gave them a considerable degree of financial security.[35]

Academia

Earlier, and upon his return to England in 1906, Wood Jones had been thrown briefly back into academic life as a demonstrator in anatomy at the London Hospital. He was then seconded to the eminent Australian anatomist Sir Grafton Elliot Smith (then Professor of Anatomy at Cairo), to work on the Archaeological Survey of Nubia. This work was extremely urgent, as the Aswan Dam was near completion and many important archaeological sites were to be flooded. The resulting joint publication, and the product of his unceasing labour, was the classic *The Archaeological Survey of Nubia: Report for 1907–8* (1910). The anthropologist, Raymond Dart, declared that Wood Jones and Elliot Smith had jointly initiated the new scientific studies of human and comparative palaeopathology.[36] In the course of this study, Wood Jones had examined the cause of death of a great number of mummies, and backed by further experiments in London, he used his observations in Egypt to devise a humane method of hanging criminals.

His relationship with Elliot Smith was to become complex. He fell out with him after the Nubia experience, although he seems not to have informed Elliot Smith, who continued to correspond with Wood Jones until his death. Some have suggested Wood Jones was unhappy with his treatment in Egypt. Another reason could have been Elliot Smith's behaviour at the meeting at the Zoological Society in 1919, discussed below.[37] Much later, on 9 November 1936, he wrote to Arthur Keith:

> I would have been very loyal to E[lliot] S[mith], for I learned to be very fond of him in Nubia – when he was at his best: away from rivals, away from controversies, away from jealousies – and in camp – he was the finest companion one could wish for. But social ambition, mainly due to his wife – intense jealousy of his fellows – intolerance – and a little devious streak, not altogether nice, made him hard to serve as a loved master ... His wife was a brilliant violinist, a charming woman – a Sydney girl, no soul, but in the place of soul, material social ambition.[38]

He did, however, contribute a chapter to Elliot Smith's biography.[39]

The nine years that followed Wood Jones's return from Egypt saw him rise through the ranks of teaching anatomists via the medical schools of Manchester University and the London and St Thomas Hospitals, culminating in his appointment as Professor of Anatomy at the Royal Free Hospital in 1915 (the London School of Medicine for Women). Female medical students were still an unknown quantity for many men, including Wood Jones:

> My time at St. Thomas's had been so completely happy that I parted from all the associations there with the greatest regret and I rather dreaded an institution in which I should have only women students and, for the most part, women colleagues ... But despite these rather gloomy forebodings, my ... connection with the Women's School was a very pleasant one.[40]

He claimed that a major reason for his success there was due to:

> the staunch friendship of Louisa Aldrich-Blake who was one of the finest women it has ever been my good fortune to have as a friend. Large of frame, large of mind, there was that about her that naturally inspired unquestioning confidence in her absolute reliability ... Other members of the staff and of

> the governing body had at that time far more public acclaim as pioneers in the medical education of women, but I fancy that no woman did more solid or more lasting work in establishing the reputation of the medical woman than Dame Louisa Aldrich-Blake.[41]

During this period, and as a sign of his rise to eminence, he was invited to deliver a number of the Arris and Gale annual lectures in the period 1914–19. Two important features that distinguish this period before his move into the army in 1918, were his development of the radical arboreal theory of humanity's evolution (crystallised in the publication of *Arboreal Man* in 1916), and his growing reputation as a teacher and public speaker.

The orator

The selection of Wood Jones to deliver the Joseph Bancroft Memorial Lecture was well chosen, as it was perhaps as an orator that Wood Jones made his most indelible mark. It also seems as though being a public figure was a prerequisite to his position as Professor of Anatomy in Melbourne in the first eighty years of the University. Wood Jones certainly proved to be a worthy successor to Halford, Allen and Berry.

In March 1935, Wood Jones gave the Syme Oration at the opening of the building erected in the centre of Melbourne to house the Royal Australasian College of Surgeons. Among the packed audience sat the Antarctic explorer, geologist, historian, and Cambridge don, Raymond Priestley. He was present in his capacity as the recently installed and first full-time Vice-Chancellor of Melbourne University. Priestley wrote:

> Wood Jones's oration was very impressive, as I had been given to expect. He spoke for over an hour on the History of Surgery, mainly upon the Master Surgeons between 1500 and 1850, before the general use of anaesthetics ... without a note, in carefully chosen language, slowly and dramatically, and yet without artificiality. All the time he was working towards a main point ... It was altogether a lesson to me in the art of lecturing, and one that I hope may not get altogether lost. I must take any opportunity I get to attend any public lecture that he may give.[42]

Wood Jones's weekly lecture in Anatomy was invariably a full house, with many sitting or standing in the aisles. The audience was made up not only of medical students but also interested parties from the University and the City.[43] His public lectures were invariably given to large audiences in big theatres. His valedictory lecture in Melbourne was booked out at the Melbourne Town Hall and was broadcast live on many national radio stations.[44] Fortunately the bulk of his lectures have survived, as he published them as essay collections.

Wood Jones first made his mark as an outstanding lecturer at the relatively late age of thirty-six, upon delivering a lecture to the Royal Zoological Society in 1915 on 'The influence of the arboreal habit in the evolution of the reproductive system.' For the layman this was an innocuous enough sounding title, but it was a topic that was to lead to great dispute in the anthropological world. In his memoir of Wood Jones as a Fellow of the Royal Society, Sir Wilfrid Le Gros Clark recounted the secret to his spellbinding public performances:

> he was equipped with unusual powers of memorisation, and the present writer has been told by a member of one of his audiences, who on that occasion had the prepared script of the lecture before him, that the latter was delivered by Wood Jones from memory with an absolutely word-by-word accuracy.[45]

Le Gros Clark added that 'curiously enough, although he was undoubtedly a pre-eminent lecturer, he was never outstanding in spontaneous public discussion'.[46] This was not, however, the experience of those students in Melbourne whom he anointed as worthy of his notice. His successor to the Chair of Anatomy, Sydney Sunderland, wrote:

> He established remarkable rapport with younger students, as well as with senior medical personnel. He had an open house on a Wednesday evening for his prosectors which were wonderful occasions. I never missed one, and there were many others. I met artists – I ultimately met [Justus] Jorgensen of the [Max] Meldrum School who'd painted Berry (I didn't have the courage to ask him about Berry's portrait as there had been quite a row about it.) I met musicians, writers, Percy Leason – the great cartoonist who went to New York and established fame and fortune.[47]

James Guest, a student of Wood Jones, and later a prominent Melbourne surgeon and teacher in the Anatomy Department, concurs with this portrait of the man. He remembers Wood Jones during McCoy Society field trips lying on his camp bed, incessantly smoking and entertaining groups of students with his conversation.[48] He certainly was an outstanding teacher. Elliot Smith wrote to Wood Jones in 1923: 'London badly wants you. For the dull weight of the other London schools of anatomy is a heavy incubus on the back of progress.'[49]

Arboreal Man

Wood Jones's arboreal theory of humanity's origins was a typical product of his quirky, anti-establishment views. Always stubborn, he was to hold to his views for the rest of his life. The beginning of the twentieth century saw a flurry of theories concerning the origin of the human species. The competing camps agreed upon the general narrative of an early primate-like creature that lived in the trees; that left the trees and developed bipedalism; and showed increasing intelligence (by developing a larger brain). No consensus could be reached, however, concerning the order in which to place these developments or the timeframe in which they occurred. One group argued that the brain grew large first and then it became bipedal, while the others argued that bipedalism encouraged brain growth. Darwin had argued that mankind arose relatively late from an intelligent bipedal ape in Africa. This offended the sensibilities of many, including Wood Jones, and it was his arboreal hypothesis (initially developed with Elliot Smith) that was the most anti-Darwinian of all competing theories. In 1919 at a meeting of the Zoological Society in London, Wood Jones proposed that so many features of ape and monkey anatomy were so atypical of *Homo sapiens* that, in fact, they must have separated early in evolutionary terms. He proposed that humanity was descended from a creature similar to the modern tarsier, an extremely unusual, small creature and the only exclusively carniverous primate. The tarsier is nocturnal, thus having large eyes, and inhabits the region of Malaysia. Wood Jones therefore posited Asia rather than Africa as the cradle of humanity. Also, instead of current estimates of the ancestors of *Homo sapiens* separating off from the apes some twenty to thirty million years ago (a view held by Arthur Keith), Wood Jones's theory placed the separation to approximately fifty to sixty million years. Even more recent studies in molecular biology seem to indicate a much later separation of the various hominids (around five million years) and the place of origin to be Africa, both hypothesised by Darwin.

His version of the arboreal theory posited that life in the trees maximised steady evolutionary progress. Natural selection in trees produced improved vision, better hand eye co-ordination, less reliance on the use of scent and higher intelligence. As for movement, trees discouraged quadrupedal motion and helped differentiate the tasks of feet and hands.[50] Tarsiers illustrate many of these qualities without demonstrating the 'evolutionary success' like that of modern humans that Wood Jones would associate with them. The two significant problems with the Tarsian Hypothesis were the absence of any substantive evidence of such a genealogy in the fossil record and the argument that, given the number of different tree dwellers, humanity alone evolved in such a manner.

His Tarsian Hypothesis was attractive to the general public as it was based on the belief that an upright, walking, intelligent human was the natural consequence of evolution without natural selection and was removed from that of apes and monkeys. As an idealist (humanity has a vital spark or soul) he believed in a teleological evolution, that is, that evolution proceeded towards a definite conclusion instead of following a random pattern. Wood Jones saw no problem with the lack of the evolution of other similar arboreal primates, as the 'soul' of humanity distinguished it from the rest of earth's inhabitants. He had more difficulty convincing his fellow British anthropologists, however, who attended the meeting of the Zoological Society of London in 1919, as one by one, his colleagues stood and denounced his theory. Most gallingly and without warning (having told the audience at length of the similarities the tarsier shared with other primates including humanity), Elliot Smith took a sharp U-turn, exclaiming:

> The ancestors of *Tarsius*, in fact, fell out of the race for intellectual supremacy in early Eocene times and ceased to cultivate their cerebral organs. Eventually, like the Lemurs, they had to adopt nocturnal habits to avoid the risk of extinction.[51]

Rubbing salt into the wound, he remarked of Wood Jones's argument that 'by means of such special pleading a case might be stated for the view that the whale was a fish, if care were taken to omit all reference to the mammalian characters of the Cetacea.'[52] Elliot Smith no longer had Wood Jones's loyalty, and the other attacks on Wood Jones were savage. P Chalmers Mitchell stated that Wood Jones:

> was more reticent tonight than in his addresses to more popular audiences. He attacked Darwinian evolutionists, and Huxley in particular, on the supposition that they believed genealogical trees to be linear, that a higher group took origin from the highest members from a lower group. It was a strange misreading of familiar evidence. Were he to consult Huxley's Essay on "Man's Place in Nature", or any general statement of the case for Evolution, as, for instance, the Article under the heading in the "Encyclopedia Britannica", he would see that he was attacking a bogey that does not exist.[53]

Another critic, Professor McBride, commented that Wood Jones had made bold public statements on this topic 'from which [he] had rather shrunk during the discussion, but to which he had liberally committed himself in recent books published by him' and he believed that 'it followed that the

The head of *tarsius spectrum* by Professor Frederic Wood Jones in *Hallmarks of Mankind*, 1948.

phrase "Arboreal Man" was a contradiction in terms.'[54] A modern commentator has written that:

> Wood Jones and his tarsioid theory of human origins never recovered from the scholarly drubbing they absorbed that day. Wood Jones exiled himself to Australia, kindled an academic interest in marsupials, and soon abandoned his research on primate evolution altogether.[55]

This claim is far from true. Wood Jones published widely on the Tarsian Hypothesis, the summation being *Man's Place Amongst the Mammals* (1929).[56] He also continued to inspire active criticism from important anthropologists, many of whom were far less critical than those who attended the meeting in 1919.[57] The leader of the Darwinian school, William King Gregory, in *Man's Place Amongst the Anthropoids* (1934), responded to Wood Jones's attacks, while also acknowledging that he was a 'brilliant anatomist'.[58] Francis M Ashley-Montagu, later known (due to a host of achievements including the creation of 'UNESCO man' after the Second World War), as the polymath, Ashley Montagu,[59] wrote a twenty-seven page review essay contradicting the Tarsian Hypothesis,[60] in which he praised Wood Jones's 'commendable unwillingness to accept the theories of others at their face value.'[61] He concluded that if 'Professor Wood Jones is able to bring forth any evidence seriously challenging the theory of Man's descent to which such an inference may logically lead, I have yet to see it.'[62] Without going into technical detail, and although generally not accepted, Wood Jones's theory was not without merit, as evidenced by modern interest in the Tarsian Hypothesis.[63]

He certainly did not exile himself to Australia due to lack of opportunity in Britain or the United States. Over almost two decades in Australia, he was offered, and rejected, numerous prestigious posts in England. Eventually in 1938, he accepted the Chair of Anatomy at Manchester University, and finally the William Collins Chair of Comparative Anatomy at The Royal College of Surgeons of England. A certain amount of bitterness, however, concerning the attacks on his theory remained. He wrote to Arthur Keith in 1929: 'if Elliot Smith had refrained from his nasty methods with me I would not have been at the trouble to hunt out his inconsistencies in his own special field. The same applies to [William King] Gregory (the leader of the Darwinians) – a shallow fellow.'[64]

Arboreal Man (aka Wood Jones) Descending from the Trees.
Cartoon in *Speculum*, July 1930, p.26.

The First World War

Early in 1918 Wood Jones took leave from his Chair at the Royal Free Hospital and joined the Royal Army Medical Corps as Captain. He had attempted unsuccessfully to enlist for a number of years before this.[65] He worked at the Military Hospital, Shepherd's Bush, London, with the team of Robert Jones, a formidable and highly influential orthopaedic surgeon. The lectures he gave on the hand during this period were published as *The Principles of Anatomy as Seen in the Hand* (1920). A highly influential text, and described by Keith as 'your great "hand" book'[66], it was used widely and served as an important reason for a group of leading Melbourne surgeons securing his appointment in 1929.

In his work on atolls, and later on the origins of humanity, Wood Jones had developed a strong dissenting position from the Darwinian heritage. His experience of the First World War strengthened his stance, especially as he rejected natural selection or the 'survival of the fittest', ideas he saw as contributors to the folly and slaughter in the battlefields of France.[67] In fact, he had cast himself adrift from Darwin and his followers, especially the German, Ernst Haeckel. In 1938 and shortly after leaving Melbourne, he wrote:

> I have many times called attention to the part that appears to have been played by the wide and unthinking acceptance of all the worst connotations of Charles Darwin's phrase 'the struggle for existence' and Herbert Spencer's 'survival of the fittest'... [that] the Darwinian survival thesis has gone far towards producing the sinister theories that have let loose the present demons of bloodshed and destruction is not to be doubted.[68]

Australia and the Pacific 1920–38

In 1919 Wood Jones was appointed the Elder Professor of Anatomy at the University of Adelaide. He arrived with a Malay wife, five adopted children and two pet monkeys. One of his monkeys, Chico, when punished 'would take a handkerchief and appear to sob bitterly as she held it over her face.'[69] It must have seemed as if all Anatomy professors were eccentric in Adelaide, as his predecessor, Archibald Watson, rode a motorbike around the dissecting room—one of a number of bizarre habits.[70] Race, however, was another matter in the interwar years in Australia. The Czech-German Communist, Egon Erwin Kisch, observed the White Australia Policy from close quarters during his visit in 1934. He wrote that as a consequence of the White Australia Policy, 'Australia consists of 96 per cent white skinned, "British born" inhabitants, and about this the Labour leaders are just as proud as the bourgeois Nationalists.'[71] There is little evidence that remains of Gertrude's reception into Australian society, other than the telling fact that she never accompanied Wood Jones on public occasions.[72]

Shortly after his arrival in Adelaide, Wood Jones received a letter offering advice from his counterpart in Melbourne, Richard Berry. Berry welcomed him to 'this remote country' and then attempted to cheer him up:

> Your [Anatomy] Department in Adelaide is the worst of the three, those in Melbourne and Sydney being far beyond it in every way, however, your Department is a good deal better than was mine in Melbourne when I first came to it 15 years ago ... Sydney is easily the best University in Australia ... as we work round the Australian coast the Universities progressively get worse. Your own Department isn't a Department at all. It is just a mortuary where students dissect dead bodies.[73]

He also warned him against criticising his new country: 'The native born Australian, of which I am not one, thank God, is peculiarly sensitive to criticism, and my advice, if followed will make your job a little easier, for you are up against it all right.'[74]

In fact Wood Jones did not find the Anatomy Department as derelict as the one described by Berry. Although the medical school was small, virtually the whole of the student body were returned soldiers, making discipline redundant and teaching and supervision light. He wrote:

> with willing help from some of the younger surgeons, the Department was a very pleasant place in which to work. I had a good workroom of my own and was fortunate in that the University is next door to the State Museums and Library.[75]

It was in Australia that Wood Jones would refine his version of philosophical idealism in opposition to the materialism of the Darwinian world view. He wrote to Arthur Keith in 1927:

> Do you still fight? I ask this because I get more and more certain ... of the early differentiation of Homo, *before* the anthropoid-monkey development and ... more and more convinced of the certainty of the reality of the inheritance of (even trivial) "acquired" characters [Lamarckism].[76]

Wood Jones was a confirmed anti-Darwinian in every respect, and would concentrate much of his efforts in the two decades spent in Australia and the Pacific Region (from 1919), on developing his Lamarckian theories of evolution. Australian animals were to provide much ammunition for his battle with the Darwinians. One example he used was that the hair follicles in marsupials grew in different directions from the embryonic stage when compared to other mammals. Wood Jones argued that their different grooming habits had changed their genetic information. Others claimed they merely learnt to groom hair that had random genetic variation, and both arguments held firm. Although Lamarckism was in decline throughout the twentieth century, there have been recent signs of its resurrection.[77] Wood Jones's Lamarckism seems as much a product of his deep dislike of social Darwinism as of any disinterested scientific curiosity. For example, he championed the (later to be disgraced) Russian Lamarckian geneticist T D Lysenko.[78]

The real attraction for Wood Jones in Adelaide was the strange flora and fauna. This was his first protracted stay in Australia, and it was spent on numerous excursions to remote arid parts of the state as well as his special passion, the numerous islands off the South Australian coast. These trips took up all his spare time, involved much discomfort and were fraught with danger, especially snakes. He particularly loved Kangaroo Island, the third largest of Australia's islands (about one hundred and forty kilometres long and forty kilometres wide), close to Adelaide and visible from the Fleurieu Peninsula. With a dry temperate climate, the island teemed with native flora and fauna. Wood Jones was successful in making it a reserve and he introduced numerous mainland species with the aim of preserving and protecting them from the degradation of development. In 1926 when Wood Jones contemplated leaving Adelaide, he wrote to Arthur Keith:

> The thing that I am most sorry to leave is the Fauna and Flora Reserve on Kangaroo Island. I joined the Board in 1921 and then we had a few shillings credit and had done nothing in the way of development. Today we have 200 square miles of bush patrolled by a resident ranger; we have a fresh grant of 16,000 acres containing caves full of bones of an extinct fauna ... The Reserve is teeming with marsupials and monotremes and we have introduced several forms such as ... the Emu from the mainland ... I am sorry to leave all this for there is no place I love more than that Reserve, and I have just returned from a delightful week spent there with the wife and the girls. It is a beautiful place and although I cannot say how many visits I have made and how many collections I have got together there, there is still much to do.[79]

Melbourne was to give him the opportunity for the continuation of his work on island ecologies when he founded the McCoy Society. The results of his efforts in Adelaide are contained in his seminal illustrated work, *The Mammals of South Australia* (1923–5). His skill in drawing, admired by his peers at the time, can be enjoyed in this work. He was elected a fellow of the Royal Society in 1925.

The other great theme of his time in Australia was the plight of the Aborigines. This he would also take up with great vigour in Melbourne in the 1930s. He wrote in 1926:

> I believe that few people, even city dwellers in Australia, realise the deplorable condition, and the disgraceful treatment of the Australian native. I, for one, have learned to like, and to admire the aptitudes of, the native; and he is by no means the lowest of the low as he is depicted in almost every published account of him.[80]

Part of his solution was to establish the Adelaide Anthropological Society made up of 'a splendid little band of younger men ... all working to obtain a sympathetic understanding of the blackfellow.'[81] As well as raising the problem publicly and in academic circles, Wood Jones applied for and secured a Rockefeller grant of $20,000 per annum, as well as scholarships to study the anthropology of the Aborigine.[82] He was to be bitterly disappointed with the way in which the money was spent. On reaching Hawaii he discovered that much of the grant was to be spent in Sydney and not on field research, and Adelaide University would receive virtually nothing. He was bitter and wrote to Arthur Keith:

Bilby by Wood Jones in *The Mammals of South Australia*, 1923.

> the money that should go towards inquiry into that most interesting of vanishing human forms is being spent in Sydney and Sydney only – on administration and teaching ... I left a circle in Adelaide – but what does Sydney care for them? Australia is politically rotten with interstate jealousies, but what has science to do with this? It is a sad waste of Mr Rockefeller's dollars and I sometimes feel sorry that I was instrumental in getting them into Australia at all.[83]

Wood Jones clearly did not agree with Berry's praise of Anatomy and Anthropology at Sydney University, and he declined the offer of a Chair in Sydney in 1925.[84] It may well have been due to his strained relationship with Elliot Smith, a Sydney graduate and behind the scenes powerbroker in Sydney appointments. Also, when contemplating a move to take up the Chair at St Bartholomew's Hospital in London in 1923, he discussed his possible replacements with Arthur Keith:

> if they are typical Sydney graduates, I do not want them to fill my place. Indeed I do not want E[lliot] S[mith] to have any influence in appointing my successors here. The students are now keen, with good ideals, and good discipline.[85]

He maintained this animosity in a children's poem, 'The Sandpiper', in his book of children's poems *Sea Birds Simplified* (1934), published during his Melbourne sojourn, and written and illustrated for his grandson, Bruce Axel Clunies Ross:

> Where ripples meet the sand
> The long-legged waders stand,
> Or hunt among the jetsam for their dinner.
> All waders are akin:
> Their bills are long and thin.
> Their elongated legs are even thinner.
> ...
>
> A man in Sydney—blessed be his memory—
> Confused these birds with some form of emery.[86]

Wood Jones's drawing of a sandpiper for his book of poetry for children *Sea Birds Simplified*, 1934

In 1927 Wood Jones was approached by the Rockefeller Foundation to be the inaugural Professor of Physical Anthropology at the University of Hawaii. Due to the isolation of Adelaide and the larger salary on offer, Wood Jones accepted. He left his work on Australia unfinished but wrote at the time: 'I have been here seven years, and I have collected material upon which I can work anywhere in the world, so my Australian work is not ended. I still have marsupial anatomy.'[87] He could not have guessed at this time that he would return within a few years.

Honolulu and race

In Hawaii the main attraction was the opportunity to study racial anatomy. I have discussed in the previous chapter the immense intellectual and political status accorded Physical Anthropology from the middle of the nineteenth century until 1939. In Honolulu Wood Jones teamed up with Stanley Porteus, yet another Melbourne connection. Porteus had been the foundation headmaster at the School for Mental Deficients in Bell Street, Fitzroy, and for many years until 1919, he was Berry's assistant craniometrist in the Anatomy Department. Porteus had risen to Chair of Psychology in Hawaii without ever having gained a degree. His fascinating biography cannot be recounted here, but he was a committed eugenicist, unlike Wood Jones.[88] They collaborated on *The Matrix of the Mind* (1928) which received lukewarm reviews.[89] Hawaii, however, fell below Wood Jones's expectations; the American culture left him cold. He wrote to Arthur Keith as early as 1928, 'the more I see of American work, the worse I think of it. There is this damned idea of "production" about it. Scientific work for them is like motor cars – you can speed it up and have mass production.'[90] It was while he was in such a mood that he received an invitation to replace Richard Berry as Professor of Anatomy in Melbourne. At first he refused because the drop in salary was too great (from the equivalent of £1,500 in Honolulu to £1,300 and a pension),[91] but a group of wealthy Melburnians led by the surgeon, Alan Newton, managed to raise the necessary funds to increase the salary by £200 per annum,[92] and Wood Jones accepted.

Frederic Wood Jones versus Richard Berry

Wood Jones faced various problems upon his arrival in Melbourne in 1930. The Royal Australasian College of Surgeons (formerly the College of Surgeons of Australasia) had been established in 1927 in the face of considerable animosity from the native British Medical Association.[93] Wood Jones had been lured to Melbourne with extra money supplied by surgeons and he felt obliged to work hard to support the College in its foundation years.[94] Another discouraging factor was the Depression, with its accompanying financial stringencies, and its effects were being felt in Melbourne just as Wood Jones arrived. And finally, the extra money offered to him as an incentive to accept the Chair infuriated Richard Berry.

This, as well as the manner in which Berry departed, created a poisonous atmosphere for Wood Jones and he wrote:

> Berry was a funny character. His going was rather bitter and evidently he was very much enraged at my election – or rather invitation. He has made all the difficulties he could for me – both by a legacy of words and deeds. He must have been designed on small lines, to have thought it worthwhile to do such small things for my discomfiture.[95]

It did not help that Berry had wanted his loyal deputy, Colin McKenzie, to follow him and Wood Jones and McKenzie soon fell out. After a year had passed Wood Jones felt he had finally begun to stamp himself on the Department, and he wrote to Arthur Keith:

> I have now got my own men (the Berry atmosphere or taint, depending on the way you look at it has worn off, and they are my own brand). With a generation of my own I feel far more comfortable and more enthusiastic – and we are going to work ... I am so sick of seeing in Brass or in Granite. "Berry F.R.S." in my Department.[96]

Fortunately for Wood Jones, this bitter legacy did not last. Aside from teaching, the three important themes of Wood Jones's Melbourne period were the Aboriginal question; the crystallisation of his Lamarckism alternative to Darwin's theory of natural selection (published in *Habit and Heritage* in 1943); and his synthesis of all the disparate strands of his studies under a unifying theory of life.

The Aboriginal question

Richard Berry's measurements of adult Aboriginal skulls convinced him that indigenous Australians had the brain size and thus mentality of a thirteen-year-old Anglo-Saxon child, as did the bushranger Ned Kelly.[97] Wood Jones disagreed violently. He wrote in *Unscientific Excursions* (a collection of essays, many of which were originally published in the Melbourne *Argus*) that 'the worshippers of the large brain claim far too much ... and many of their claims are false and that is why I range myself in opposition to them'.[98] He pointed out that Neanderthals had a much

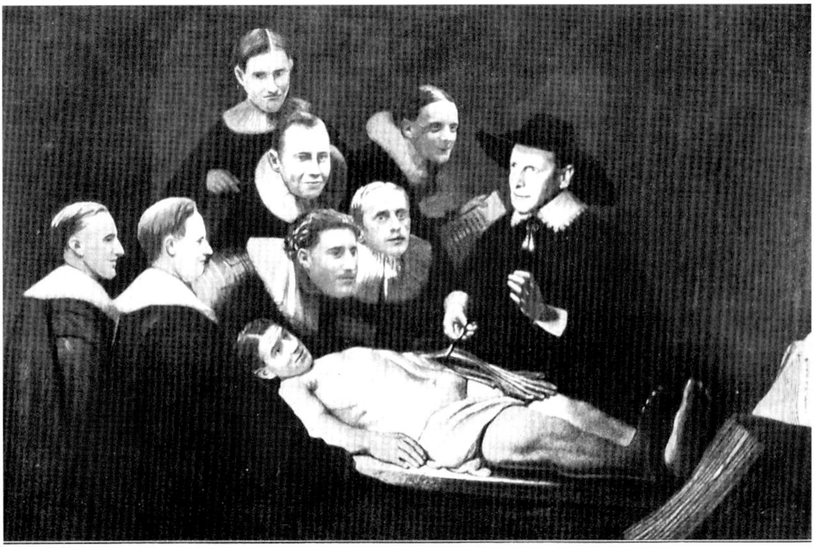

Wood Jones's Anatomy Lesson to Melbourne Students
(with apologies to Rembrandt) in *Speculum* July 1935, p.26.

larger brain than modern individuals. In relation to the Australian Aborigines he remarked:

> the Rev[erend] Mr Horton wrote of the natives of Tasmania: "Indeed the shape of their bodies is almost the only mark by which one can recognise them as fellow-men; ... they are a race of beings altogether distinct from ourselves, and class them among the inferior species of irrational animals." ... I should have dearly loved to tell him that many of the natives of Tasmania had brains far larger than his; and, unless his head was unusually large, the right would have been on my side ... Truly, modern civilized man is not supreme in this matter of big brains, for his forgotten ancestors and his despised, pigmented brother may both surpass him in this regard.[99]

His first argument with contemporary orthodoxy was that he disputed the notion of the Aboriginal as inferior to the white race. In three broadcasts on the Melbourne radio station 3LO on the 5, 11 and 19 July 1934, he stated:

> Intelligence is an attribute that permits an individual to adjust itself correctly to the changing demands of its environment, and it is useless to attempt to

> establish a simple criterion, by which the intelligence of a nomadic stone-age hunter and of a white politician may be measured and contrasted.[100]

Nor according to Wood Jones could the Aboriginal be seen to be physically inferior as 'it is safe to say that no more beautiful balanced human figures than those of the native in the prime of his life could be found among any race.'[101] His explanation of the evolution of Aboriginal society and culture is strikingly similar to the explanation first offered in 1997 by Jared Diamond in his popular and influential book and television series, *Guns, Germs and Steel: the Fates of Human Societies*.[102] Preceding Diamond by some sixty years, Wood Jones argued that the development of modern civilisation is only possible with agriculture and the leisure that follows from the production of excess food. It was therefore impossible for the nomadic Aboriginal to settle and build cities, as there were no species of fauna or flora in Australia that were capable of domestication:

> Indeed the white colonist has made no more conservative use of the native animals and plants of Australia than did the aborigine, for in no single instance has he adapted to his needs any animal or plant indigenous to Australia. All that can be said for the white colonist (in this regard) is that by a thoroughly vandalistic policy he has made a profit out of the destruction of native fauna and flora.[103]

Conclusion

Wood Jones resigned from the Chair of Anatomy in Melbourne in 1937. Rather than follow his initial plan to withdraw to the English countryside, he accepted the offer of the Chair in Manchester, and in 1945 he was offered the position of Conservator of the bomb damaged Hunterian Collection in The Royal College of Surgeons of England—an appropriate conclusion to the career of one of John Hunter's most avid disciples. His departure from Melbourne was typically controversial. His valedictory lecture on 9 December 1937 was given to a packed Melbourne Town Hall, and broadcast live on radio across the nation. In his address 'Universities and Life' he attempted to present his vision for an ideal university. Freedom of thought and the pursuit of excellence were, naturally enough, essential. Although a vitalist—believing that humanity was not just the product of various inorganic chemicals, but rather possessed of a vital spirit or soul —Wood Jones was not a supporter of traditional religion. He shocked many in his audience by claiming that Christianity, in the form of the University colleges, offered nothing more than 'stultifying bondage' in the life of the University. Most particularly, according to Wood Jones, 'the division caused by sectarian issues is one of the outstanding defects of the intellectual life of Australia.'[104]

Frederic Wood Jones was a remarkable intellectual, possessed of both broad and deep knowledge that spanned many disciplines. His most important legacy, apart from his powerful influence as a teacher on a generation of physicians, surgeons, anthropologists and naturalists, was as a conscience and critical voice, always questioning any scientific theory that became inviolable law. He managed to perform this role without referral to the skills most commonly admired among scientists of his generation—those of the laboratory scientist. Although remarkably prescient, he was a man of the past in many ways; a nineteenth century naturalist in the image of Charles Darwin. As his former student and fellow member of the McCoy Society, James Guest, later wrote:

> Wood Jones was at his best in the field observing the living animal, noting its habits and habitat, dissecting specimens, drawing, making more notes and then weaving what he had learned into the broader fabric of his ideas on adaptation, evolution and unity in nature. His tools were his eyes, his hands, a scalpel, a pen, a hand lens and (to hone the scalpel blades razor sharp) a small oilstone.[105]

Wood Jones had no great love for the confines of the museum, the dissecting room or the laboratory. He wrote:

> Anatomy is not for the dissecting-room only; the anatomist is not only for the dead-house; he is for the wards, the operating theatre and the clinic, if he is to serve his university and the community to the full.[106]

His vision for the anatomist also encompassed the entirety of nature. This is perhaps best summed up in a Chinese proverb quoted by Wood Jones concerning the limitations of the laboratory scientist: 'It is useless to speak about the sea to a frog that lives in a well.'[107]

Wood Jones sailing in Spencer Gulf while on an expedition to Reevesby Island with the McCoy Society in 1936. Courtesy: James Guest.

CHAPTER SEVEN

The teaching of anatomy, 1906–39: 'our anatomy is no longer dry bone'

THE *SPECULUM* ISSUE of December 1905 reported there was 'no end of speculation'[1] as to the character and appearance of the new professor, and how he would transform the Medical School. The unsuspecting School, however, was in for a jolt. Richard Berry's appointment was to create fireworks, due as much to his volatile and driving personality, as the need for a thorough overhaul in the teaching of anatomy. Word soon got round the Department that it was 'better dead than Berried'.[2] Certainly, in principle, his arrival was welcomed, and his enthusiastic reception at the 1906 Commencement Ceremony was an expression of such goodwill.[3] After a year of teaching, the student magazine reported:

> Our anatomy is now no longer dry bone,
> Since Prof. Berry has come to expound it;
> For, in some strange poetical way of his own
> He has weaved graceful garlands around it.[4]

With the benefit of knowing those with personal recollections, Kenneth Russell wrote:

> the impact of his teaching on the students was immediate and profound. He took complete control of the discipline of the classes, his dogmatic lectures being given in a staccato manner with a rather grating voice; he laid emphasis on the practical aspects of the subject and enlivened it with a rich and devastating vocabulary.[5]

The inadequacies in teaching anatomy in Melbourne in 1906 were threefold: a lack of good curriculum material (such as wet and dry specimens, transverse sections and other subsidiary aids), the lack of adequate accommodation and unsatisfactory teaching arrangements. Berry tackled these problems with great success in the first few years of his appointment with one notable exception, the lack of well-paid teaching staff. Initially, however, his reforms met with less than universal support.

Richard Berry's first act in 1906 was decisive and true to his character. He began his reforms through the establishment of a simple but effective code of conduct in the dissecting room:

> No smoking or general conversation is permitted in the Dissecting Room. Any student creating or taking part in any disturbance whatsoever in the Dissecting Room will be severely dealt with.[6]

Though change is often eagerly anticipated, frequently the reality seems less attractive. Thus Berry's first reform prompted a plaintive reply in verse in the student magazine:

> Take away my Grey and Cunny,
> Pawn my instruments and "mike",
> ...
> For I'm going out on strike.
> Now the work has lost its flavour,
> (Though the odour's rather high),
> Since we beggars must behave our
> Selves not lark upon the sly.
> ...
> It was jolly in the old years
> When we had some liberty;
> Then we fought like valiant soldiers
> Shying meat right royally.

> Then we sometimes got a jawful,
> Or we sometimes got a fine,
> ...
> And you've gallantly attempted
> Now to keep our sport in check,
> And the meat we once emptied
> Down another fellow's neck
> Must repose at peace in dishes.[7]

After initiating a stricter regime in the dissecting room, his next act in the following year was to introduce a new system of supervision, a reform that was long overdue. This involved investing a group of the cleverest students with the title of prosector and empowering them to act as supervisors. Prosectors had existed in title, but had never had the status or power invested in them by Berry. Masculinity was also a prerequisite. In 1907 the single woman student among the top three candidates was debarred from accepting a prosectorship on account of her sex.[8] Despite this injustice, the first appointments proved to be propitious as Alan Newton, Victor Hurley and William Upjohn were to become 'a surgical triumvirate unique in the history of this school.'[9] This, however, did not satisfy everyone. In fact *Speculum* argued against the whole system 'with the justice of their appointment, with the method of their appointment, and with their usefulness.'[10] Not only were the prosectors chosen after a 'seemingly purposeless examination' where luck supposedly played a part, but also they received preferential treatment without re-examination throughout all the years of anatomy classes. Their duties and advantages were such that:

> they dissect the part of the body that is being treated in lectures; they get frequent private demonstrations from the Professor; are paid special attention from the demonstrators, and are exempt from obtaining marks on their dissections.[11]

Accordingly:

> it must follow from this that, unless the prosector is a dullard, he has every chance of learning infinitely more anatomy than one who has to dissect with

> the aid of a Cunningham [Anatomy textbook], and an occasional demonstration from one of the Professor's understudies. Thus, the exhibition in anatomy is virtually given to the best of this little admiration society.[12]

The prosectors also prepared specimens for the museum, a task dismissed as unnecessary by *Speculum*. Whatever the complaints, the system was to prove resilient, the last prosector being appointed in 2000.[13]

Berry's third act on arrival was typical of his entrepreneurial spirit and temporarily solved the problem of accommodation, when he organised funds for the rebuilding of the Anatomy School. A second floor was added to the medical building in 1907, and the dissecting room moved to the floor above, its original place being occupied by an anatomical museum. The anatomy collection was begun with some 'donations' from Allen's pathology museum.[14] The students commented:

> He's got, we accept his own warrant,
> A pretty big head for his size,
> And a power of invoking a torrent
> Of language that causes a rise ...
>
> He's got a new building to rule in
> And a lift which he lifts every day,
> Well, perhaps the lift is to fool in,
> For that's what the students all say.
> And a room that is marked the Professor has got,
> Though it may be he'd say the accent is not.[15]

Many were the jokes about the size of Berry's head, such as 'the structure is mostly head'[16], as well as references to his 'weirrd' or 'aphasic' accent halfway between Lancastrian and Scots.[17] In a fictional interview with 'Dicky' we are told:

> Picture to yourself a Cunningham, a Quain, a Morris and a Gray compressed by some marvellous force into one being; imagine further, a massive nasion, a problematic bregma, an ironical union felted with a mighty tow-like mass of independent capilli into an aesthetic angle of facial index such as might have done credit to a Wellington or a Napoleon. Dauntless dogmatism born of a

'Dicky' Berry cartoon in *Speculum*, July 1930, p.38.

constant acquaintance with agonizing anatomy, Nature had largely portrayed over his features, and bearing this in mind you have a life-like or rather a death-like mask of Dicky.[18]

Although Berry's early reforms greatly enhanced the quality of the course, the immediate intractable impediment to the satisfactory teaching of Anatomy at the University of Melbourne (continuing in the first three decades of the twentieth century) was the chronic shortage of staff. Berry's solution (apart from investing the post of prosector with some punch) was to labour tirelessly throughout his twenty-three-year tenure. And to his credit he managed a considerable research output, while also managing to play a vital part in politics both within and outside the university. He took on extra-mural activities, involving the teaching of extension subjects throughout Victoria often in areas other than medicine, for example, the French Revolution. He managed to deliver a course in anatomy of high quality to hundreds of medical and allied anatomy students.[19] Many contemporary professors lacked Berry's energy.

The shortage of staff created a string of problems for universities in Australia in the early twentieth century. As Wood Jones wrote to Arthur Keith on the former's arrival in Melbourne in 1930, when discussing the forthcoming examinations for the College of Surgeons of Australasia, 'I understand that examiners may come out from home to take primary fellowship exams here – that would be a blessing for me, for at least I should have the chance of a talk with an anatomist every now and again.'[20]

The first public forum in which Richard Berry raised the problem of the dearth of university staff was at the First Educational Congress held in Melbourne in 1912, and instigated by Berry.[21] The national efficiency movement provided the inspiration for the conference. In his speech, 'The Modern University and its Upkeep', Berry savagely attacked the state government's commitment to tertiary education.[22] Quoting statistic upon statistic when comparing Melbourne to universities in Canada, Europe, England and Japan, he demonstrated that 'the University of Melbourne is, at the present moment, the most inadequately staffed university in the world.'[23] Among the examples Berry gave were Tokio (Tokyo) University, with one professor to every thirty-five students (the Melbourne ratio was 1:66); McGill University in Montreal, with one qualified instructor to every seven students (the Melbourne ratio was 1:15);[24] Liverpool University, with the same number of students but with fifty-five professors to Melbourne's eighteen, and one hundred and ten lecturers to Melbourne's sixty. And the picture was no better in Anatomy. According to Berry, no less than three hundred and ninety-eight students engaged in the study of Anatomy for Medicine, Dentistry, Massage and Domestic Economy in Melbourne in 1911. And only one professor, three full-time assistants and one half-time assistant offered tuition. He continued: 'Of the full-time assistants, one is engaged in other work, and the remaining two have since left, and the whole weary task of training up a new number of assistants has to be recommenced.'[25] Berry then compared the Department to others, showing that Melbourne had about half the staff of a range of other departments around the world. The consequence being that 'the Melbourne Professor does extremely well if he keeps up his wicket.'[26] Perhaps even more galling was the fact that 'during the last six years ... for every £1,000 spent in Melbourne on University buildings Sydney has spent £5,000' and New Zealand, with sixty per cent of Victoria's population, spent twice as much on its University than Melbourne.[27] What did this say about Victoria and its citizens? Berry concluded:

> During the whole of my sojourn in this state I have repeatedly asked myself why is it that higher education fares so badly in Victoria, as contrasted with other states and countries? Sometimes I am told, "Oh, look at our small population; we can't afford to spend money on the University." I do look at that population, and then I glance at the Melbourne Cricket Ground and the Flemington Racecourse, which I am told are the very finest of their kind in the whole world, a statement which I believe, and which harmonizes but badly with the premise.[28]

But it was not only the lack of state spending that Berry bemoaned. He also pointed out that 'there is the almost total absence of the pious benefactor.'[29] Berry told his audience at the Congress that in Sydney the University is endowed with an annual income from its benefactions of £24,501, while Melbourne's (from corresponding sources) was only £5,600, or less than one quarter. Melbourne was without wealthy benefactors: 'The Rockefellers, Carnegies, Beits and many worldwide known benefactors of other Universities are altogether unknown in Melbourne.'[30] Berry's arguments were dismissed in the press and the editorial of the *Age* claimed:

> It is easy to point out, as Professor Berry did, that the University needs £250,000 spent upon it to bring it up to something like an ideal condition ... But the State's purse is ... not inexhaustible [and] ... success does not lie in the mere capacity to spend, but in the wise use of means.[31]

This situation would in essence not change throughout Berry's tenure, nor even during that of Wood Jones. On 21 November 1935 the Vice-Chancellor, Raymond Priestley, wrote in his diary:

> I begin to despair of the place. The medical school is supposed to be the best of the university schools ... [but I have received] reports from the Dean of the Faculty and the Professor of Anatomy to the effect that departments are hopelessly understaffed.[32]

Almost two years later the situation was no better. In a special report to Council, Priestley laid out the extent of the problem:

> In the Department of Anatomy ... we have ... only a skeleton staff. I discovered last year that for two years Professor Wood Jones had been unable to provide any classes in Dental Anatomy or to admit the dental students to the medical lectures because of overcrowding. This has been remedied but at the cost of swelling a deficit already overlarge. A few weeks ago, when I was discussing with the professor a decision he had previously reached to retire at the age of 60, he told me it was out of the question for him to consider staying. At Melbourne, he said he had, for the first time in his career, found himself in such a position that he was unable to do his job to his own satisfaction. One Senior Lecturer has resigned ... .'[33]

Wood Jones did in fact resign in the same month; a decision Priestley believed would 'wake up some people'.[34] It failed to do so and Priestley resigned shortly afterwards, citing the indifference of the government in the face of requests for adequate funding. The combined effect of dual resignations did ultimately stir the government to take action.

The situation in the first three decades of the twentieth century would have been even worse if James Stewart, a Ballarat medical practitioner, had not left a substantial bequest of £25,624 to the Medical School in 1909. This allowed the Anatomy Department to use its share to appoint a full-time Stewart Lecturer and Demonstrator in Anatomy—the first being John Hubback Anderson. The bequest also allowed the fitting out of a neurology laboratory in the Anatomy Department, and further funded three lectures every second year on a subject of national importance. This was a facility that Berry later used to spread his eugenic message of the growth in numbers of so-called mental deficients.[35]

Berry's complaint about the ephemeral nature of his support staff in 1912 was another factor inhibiting the successful running of the Department. Because the pay was poor, virtually every appointment beneath the professor was accepted on the basis that the recipient would gain experience, and eventually move on to a better paid academic appointment, or work elsewhere (usually as a surgeon) to gain a more substantial financial reward. John C Eccles (later Sir), who was a demonstrator during this period, was one such individual who eventually won a Nobel Prize in Medicine. Also, the salary of the Stewart Lecturer (a full-time appointment) was insufficient to retain an individual for any length of time. The Stewart lecturers were normally busy surgeons, and usually returned to more lucrative surgical practice. Therefore most accepted the position for no more than one year and Berry's frustration mounted. It also meant that the only significant, full-time academic in the Department was the professor.

An audit of appointments to the Department from the first Stewart Lecturer in 1909 until 1924, supports Berry's complaints about the lack of adequate support and the itinerant nature of the support he received, especially given the large increase in student numbers after the First World War. After just two years in 1911, Anderson relinquished the Stewart lectureship to (later Sir) William George Dismore Upjohn, who became a demonstrator, and in 1915, a part-time lecturer in Surface and Applied Anatomy, and then Stewart Lecturer in Surgery.[36] Berry had asked the Council for a 'second officer' in Anatomy without success, and in 1912 one of the major connections with the pre-Berry era, George Rennie, senior lecturer in Anatomy, died. When Berry was in Europe in 1910, Rennie had stood in as Acting Professor. And with Colin McKenzie, he had met Berry on the boat and issued the dire warning about the state of anatomy in the University. McKenzie had continued to assist in the Department in a part-time fashion for many years. Another ongoing part-time staff member was the surgeon (later Sir) Alan Newton, lecturer in Regional and Applied Anatomy until 1928, and significant in many roles in the teaching of surgery until his death in 1949. It was Newton, of course, who was responsible for the appointment of Wood Jones.

In 1915 Upjohn resigned and in the following eight years five individuals would fill the Stewart lectureship. They were Athol S M Tymms (1915) and Charles Roy Lister (1916–7), and the position was vacant in 1918.[37] F G Middleton was Stewart Lecturer from 1919 to 1921, followed by Ernest W Chenoweth in 1922 and John Sydney Green in 1923. And John Anderson was made the first permanent senior lecturer.

Anderson, however, returned to Britain in 1923 to a better paid position. In 1924 Henry Searby replaced him, but he too left in the following year, and the position was taken up by Albert E Coates who remained for two years.[38] Coates was to have an ongoing relationship with the Department and the University for many years, as Acting Professor in Berry's absence (something he was immensely proud of), and he became quite a Melbourne figure. He was once asked what he thought of physicians, and despite his great friendship with Leslie Hurley (perhaps Melbourne's most prominent physician) and after some consideration, he remarked he 'probably thought they were here to stay.' Such was his confidence in surgery, he was said to cry out on the admission of a patient to Emergency, 'we'll operate', followed by, 'what's wrong with the patient?'[39]

For many years from the mid 1920s the extent of the full-time staff in the Department was, apart from Berry, the Stewart Lecturer and Lecturer in Histology, Leslie E Hurley. Hurley was the younger brother of one of the first prosectors, subsequently distinguished surgeon and Stewart Lecturer in Surgery, Sir Victor Hurley. In the Anatomy Department, Leslie Hurley had been appointed lecturer and demonstrator in Histology in 1918 and was promoted to senior lecturer in 1923.[40] As mentioned, he was one of Melbourne's most popular and successful consulting physicians, and finally relinquished his position at the University in 1933 due to the pressures of private practice. Compounding the staffing problem in 1923, Allen suffered a stroke and to Berry's burden was added the responsibility of the Deanship. As most staff only stayed for a brief period of time it would have been impossible to run the School without the work of part-time and honorary staff.[41]

The curriculum

As in the half century before Berry's arrival in 1906, anatomy continued to be at the core of the medical degree, especially the newly separated pre-clinical years. While one of the main purposes of the nineteenth century anatomy course was to establish the status that separated university trained doctors from highly competitive alternative practitioners, this proved to be less important in the twentieth century. With the new confidence brought about by the victory of the British Medical Association (BMA) in destroying alternative medical practices as effective opponents, the detailed study of anatomy had been vindicated. What was new was the greatly enhanced status of surgery that was able to cure in ways that could not have been imagined before. Anatomy provided the essential training for the surgeon, but its new role proved to be controversial in the early decades of the century, as many claimed competency as surgeons, having only studied anatomy for a few years.

In 1929 Berry outlined the pre-clinical course in Anatomy to be taught in the second and third years:

> Students in the dissecting room are divided into small groups under the charge of a part-time demonstrator who attends two hours a day for two days a week. With him they discuss their difficulties and from him they receive their tuition [the students dissected five mornings a week during the Anatomy course.[42]

This was backed up with a weekly 'didactic' lecture for the whole junior class 'by means of lantern slides and diagrams' but without dissected parts because of the size of the theatre. Also for one hour each week, the students were 'introduced by the didactic lecture, lantern, and microscope to general principles underlying the cell as the unit of construction ... and to histological structure generally.' The third non-dissecting hour of the week was given over to embryology and was not compulsory.[43] The examination at the end of the first year proved crucial because this would be the time when Berry would choose his prosectors; a method not universally favoured by the student body. According to Berry the examination had:

> a simple and essentially practical objective, for the specimens, macroscopic, osteological, microscopic, and the like, are set out in front of the examinee, and he [sic] is given five minutes in which to write an answer to the question asked. Provided he can prove himself beyond the standard of average mediocrity, he is not only allowed to "pass", but is exempt from all further examination in Anatomy in this year of study.[44]

It was from among the best in this exam that Berry chose the prosectors and other members of the Honours class. Being an elite they were provided with the 'direct supervision of the Head of the Department

[Berry] and with him they themselves discuss advanced biological and neurological problems, write essays, examine literature, undertake special dissections, and so forth.'[45]

In the second year of anatomy, which occurred in the third year of the medical course, the students were introduced to 'a very complete form of neurology, the basal idea of which is that minute structure correlated with function is the key to the phenomena of mind and its aberrations, as well as to most other phenomena of health and disease in all living biological forms.'[46]

Histology

Histology (the microscopic study of tissues) had become a serious academic pursuit from the middle of the nineteenth century, when microscopes became reliably effective and widely available. The understanding that much disease existed in the tissues was an important part of the biomedical revolution. In the English tradition, it was viewed as being part of the study of Physiology, thus Halford had first taught this subject in the Medical School from about 1885. Later on, the Reverend Walter Fielder (although medically unqualified), together with John F Wilkinson, took over the delivery of the course for two afternoons each week in second year in the first two terms, with some practical work in third year along with honours classes.[47] In 1909 among revisions to the curriculum, Professor Osborne of Physiology suggested that Histology be transferred to the Anatomy Department, as the German practice of teaching the subject with anatomy was gaining currency worldwide. And Fielder was promoted to lecturer.[48] This was a most significant event in the history of the Department, which is reflected in the current identity of the Department as that of Anatomy and Cell Biology. Other anatomy departments around the world that did not incorporate Histology, proved to be less flexible in meeting the demands of evolving curricula.

Whether Osborne wanted to get rid of Histology due to problems with Fielder's teaching is unknown. From 1909, however, there were many complaints from students concerning the course. In December of that year *Speculum* protested about 'the farcical methods pursued by the Demonstrator of Histology' complaining about the disorganised nature of the examination as well as 'threats to all students before the exam.'[49]

Also, complaints were made about 'the lamentable lack of teaching in this quite important subject'.[50] Concerning the state of affairs in Histology:

> Indeed, it has been a byword in the University for so long that there appears little hope of reform. Still, let us lodge our protest, vain though it be! ... the students have a right to expect that such a course should be entrusted to the care of a capable teacher ... we ask that a competent and responsible teacher be appointed ... .[51]

In 1912 Berry arranged for some assistance for Fielder. Yet the following year *Speculum* asked 'How long will it be before Histology is taken as a serious branch of Anatomy teaching?'[52] The situation went from bad to worse in 1913 when Fielder died and the Council dithered about his replacement. *Speculum* described the situation as a 'pantomimic farce' after the appointment of the third teacher in eight months.[53] In 1914 Gilbert Lamble, having been appointed as the first Stewart Lecturer in Pathology in 1912, and briefly as Acting Professor of Pathology, was appointed to teach Histology. He made a valiant and successful effort to re-establish the credentials of the course by making it more in line with medical studies or, as it was remarked 'a course of human histology and an effective preparation for work in pathological histology.'[54] He retired due to ill health in 1916 and died the following year. Leslie Everton Hurley replaced him, subsequently being appointed senior lecturer in 1923. Hurley seems to have continued the good start made by Lamble, delivering a successful course until he resigned in 1933. After Hurley's resignation, Edward Ford was appointed lecturer in Histology in 1934, and was promoted to senior lecturer in 1935 before he resigned in 1937 (the same year as Wood Jones) to take up a position in the School of Public Health and Tropical Medicine at the University of Sydney because his interest lay in the study of malaria. Ford eventually replaced Harvey Sutton (also a Melbourne graduate and one of the first Medical Officers to be appointed to the Victorian Education Department in 1909), as Director of the Sydney School and then its Professor in 1948. Although Ford was an excellent teacher, the problem still remained of when the histology course should be introduced and taught in the curriculum. As a 'single subject' serving different masters – anatomy, physiology and pathology – it was always difficult to satisfy all, especially as there seemed to be 'a lack of cooperation' between departments.[55] For all the problems associated with the teaching of Histology, its ongoing value to the Anatomy Department was enormous.

Student numbers and free education

From 1919 to 1923 the medical course was extended to almost six years. This can be partly explained as an attempt to control student numbers, as the cessation of war had seen a significant growth in enrolments. In 1914 there had been four hundred and seven medical students, and by 1919 the numbers had increased to six hundred and sixty-four. By 1921 numbers reached a maximum of seven hundred and ninety-five. Adding to these numbers and among the dissecting students were many non-medicos, such as the artist Norman Lindsay.[56] *Speculum* was in no doubt where the blame should be placed:

> The immediate result of this sudden influx is an unsanitary overcrowding of all lecture rooms, and a criminal overworking of the staff. They cannot attend to their work properly ... responsibility for the present situation rests with the parents. It is now fashionable to send one's children to study medicine. It saves thinking.[57]

After 1911 the significant influx of British trained doctors had ceased, making way for places to be filled by local graduates. With fluctuating intakes of first year students throughout the 1920s, the ratio of doctors per capita remained remarkably stable at a little over one medical practitioner to 1,200 citizens.[58] What was not guaranteed, however, was the certainty for graduates of a secure high income. The general inability of the majority of the population to pay doctors' fees meant that without substantial private means, new graduates were forced to enter a highly competitive medical marketplace, or work for the lodges at a fixed income to survive financially. A Melbourne degree offered graduates the entrée to a profession wholly dominated by Melbourne University trained practitioners. Alternative practices had, for all intents and purposes, been eliminated as serious rivals.[59] By increasing the length of the course, and with it the fees, this had the gradual effect during the 1920s of decreasing enrolments until there were only three hundred and ninety-three medical students by 1930. There was no set quota or entrance score and it was not until the 1950s that the latter was introduced. Before the entrance score, students only needed to matriculate and have the money to pay the fees. Calls to set limits on numbers first appeared in the 1930s.

In the debate over student numbers Richard Berry made a significant contribution to the question of who should be able to attend University. As early as 1891 attempts had been made to abolish fees and thus increase the number of students who could attend University.[60] But Berry strongly disagreed. The main reason for his determined opposition to free university education was that it would only result in:

> the flooding of every profession, of every technical pursuit, and of many trades with a body of men and women who are not in the possession of sufficient brains for the task undertaken; but in time, by a process of evolution and the elimination of the unfit, that matter would right itself and the original *status quo* would be regained.[61]

The notion that University entrance should be available only to a limited number of poor students was based on the eugenic view that there were a restricted number of bright students among the poorer classes. Berry advocated that there be a limited number of scholarships for disadvantaged students. Also he had an early opportunity to influence government policy, being a member of the 1913 Committee of Inquiry that advised the government on university reform. At a public meeting he announced there were:

> probably plenty of people now carrying bricks who might have done a great deal if they had had education at the university, while there were people at the university who might as well be carrying bricks (Laughter).[62]

He had no illusions, however, that this would cause a flood of working-class graduates entering medicine. In his unpublished autobiography he described the glory of a *fin de siècle* dinner party, concluding that this was a way of life that had 'vanished to make way for the common world of the common man.'[63] To further compound their evil, Berry claimed that 'the proletariat ... would be quite capable of writing their names on the tablecloth, or of purloining some of the wine glasses.'[64] In fact the plan for scholarships proposed by the 1913 Inquiry (after a decade of dithering) as the alternative to a free university, eventually became government policy in the form of the *University Act 1923*.[65] The committee reported to Parliament that if the university were to become free:

> the inevitable tendency would be to swell the number of those who are tempted to try their fortunes in professions for which they have no aptitude, with the result of disappointment and often disaster to themselves while the State would suffer from diverting into unprofitable channels the energies of students who might have found useful and profitable careers in other occupations.[66]

The system that was put in place was designed substantially according to the system advocated by Berry, consisting of a limited loan facility and, as Berry had recommended in his speech at the 1912 Educational Congress, 'hundreds' of 'free' places.[67] In 1925 one hundred and eighty-eight students were nominated by the Education Department, nine by the University High School, one by the Railways Commissioners and thirteen by the War Bursars, making a grand total of two hundred and eleven students in a student population of 2,612 or 8.1 per cent.[68] The proportion of free places, as well as the method of allocating them, remained substantially unchanged until 1938 when a system of half fees was added.[69] If the number of free places (made available by the University of Melbourne in the decades after the *Education Act 1923*) substantially fulfilled Berry's aim to leave no bright poor child without the opportunity for a university education (Berry did not complain about the allocation and number of such places in the 1920s and 1930s), then this allocation would seem to be a measure of his belief in the limited talent pool to be found among the poorer classes.

Berry's Folly

The 1920s saw internecine battles in the University over plans to build grand buildings for the Medical School. Richard Berry's two great building projects represented a substantial investment of both his time and effort. And they involved campaigns conducted against serious opposition over extended periods. The first, and successful, project was the building of a large, stand-alone Anatomy Department. The second project was to combine the Medical School and the Melbourne Hospital at one site to create a clinical medical school.

The explosion in student numbers after the First World War was the catalyst for the anatomy building project. Of this sudden increase in students, Berry wrote in his unpublished autobiography:

> I viewed with considerable apprehension the prospect of a dissecting room only capable of accommodating two hundred students, and only designed for that as a maximum, suddenly called upon to have its walls almost burst asunder with a crowd of over five hundred. The prospect seemed to me to be highly improper, positively indecent, and of very dubious legality.[70]

In 1919, Berry decided to begin his campaign for an anatomy building by complaining to Council. Apart from the practical problems of overcrowding, he was aware that he held a particular responsibility under the *Anatomy Act 1862* as the only person with a license to run an Anatomy School, and thus being subject to penalties under the *Act*, those being three months in gaol or a fine of £100.[71] Such a responsibility was, according to Berry, an important reason as to why he disliked so much 'those "meat fights"' he had encountered upon arrival.[72]

On 11 April, and in response to his approaches to the Council, a deputation including Sir Harry Allen, Sir John Grice, Dr James Barrett, Professor Osborne and Berry visited the dissecting room to investigate Berry's claim of overcrowding. According to the students:

> For days beforehand we had been coached in our parts in order to produce the maximum effect. ... There was a crowd. It was like meat sales at the markets! ... When the committee arrived they were greeted with cat-calls, thumping of stools, and rattling of tin dishes on marble slabs. ... The gentlemen were much disturbed at the sight of our local black hole of Calcutta, and one of them, on seeing the prosectors, was overheard to murmur, "Gawd help our patients in the future."[73]

Berry claimed to be as 'astounded' by the scene in the dissecting room as were the other visitors:

> This was the first time in my life that I had ever seen *all* the students of a Practical Class, *all* present at one and the same time ... Picture if you can twenty tables long and wide enough to hold a full length recumbent human body, round which are gathered together like crows round their carrion, thirty students, white-coated, scalpel armed, looking as learned as possible with mostly borrowed spectacles, and you may dimly realise what the crowd was like and why the Council promptly sped to the government of the day for financial relief.[74]

Also in his autobiography and despite admitting that he had forewarned the students of the impending visit and had asked them all to attend, Berry complained that some members of the Council (perhaps James Barrett) later argued that he had artificially inflated the mob in the dissecting room, using non-medical students as ring ins. This charge he rejected: 'Heaven knows there was no reason to do anything quite so stupid; the facts speak for themselves.'[75] The statistics do in fact point to an overcrowded Medical School. Did Berry, however, make extra plans to ensure a successful outcome? William Preston, both Berry and Wood Jones's loyal assistant, later told Sir Sydney Sunderland (fifth Professor of Anatomy and a graduate of the Medical School) that indeed, friends of medical students from other faculties were present in the dissecting room that day, and that it was possible to identify them as '[t]hey were either rushing outside to be violently ill or they'd taken on an ashen grey or olive green colour'.[76]

Whatever Berry's tactics may have been, he triumphed and received a £60,000 grant (the final cost was £70,000) from the state government. Against the backdrop of splendid celebrations in 1923, the President of the BMA, Sir William Macewen, officially opened the new Anatomy building. This is the second appearance of the Scotsman Macewen in this history as he was the first to operate on a cerebral lesion. The interior of the building was planned by Berry. He greatly disliked the exterior, but he 'took care to see that the name of the Professor, who was primarily responsible for its design and building, was permanently recorded at the main entrance.' At the end of his life, he wondered 'is this record still where I left it?'[77] Could this have been a premonition of the controversy his name would stir up in the University at the commencement of the next millennium? It was a great victory for Berry as James Barrett vigorously opposed him in every debate concerning the size and design of the building.[78] The final size caused it to be immediately dubbed 'Berry's Folly' but it proved to be a valuable asset as it continued to provide adequate space for the growing Medical School for many decades to come.[79]

Clinical medicine

One of the significant failings of the Medical School was the long delay in inaugurating Chairs in Surgery and Medicine. Until the 1950s the Stewart Lectureships in Medicine and Surgery became de facto Chairs, albeit poorly paid and only notionally full-time. It is perhaps harsh to criticise the Medical School, as it was, yet again, the paucity of funds that impeded the foundation of clinical professorships.

It was at the close of the nineteenth century that the beginnings of significant advances in chemotherapy occurred, due to great and recent advances in the laboratory. Laboratories were then taken back into the hospitals and doctors resumed their research on the living body, thus clinical science was born. In the late 1880s universities in the United States began to get involved in teaching hospitals and set up the first clinical professorships, for example, a Professor of Medicine and a Professor of Surgery, and not solely offered in disciplines such as biochemistry. The rationale was to model clinical work as an academic discipline. The most influential model was the Johns Hopkins University Hospital, Baltimore, established in 1893. Its entire staff had trained in laboratory techniques in Germany and they wanted to apply their knowledge in the hospital wards.[80] These American schools built the discipline of clinical research in the laboratory, instead of observational research at the bedside.

Richard Berry, acutely aware of the importance of being up-to-date in the world of medical research and teaching, wanted the Melbourne Hospital that was located in the city to be moved to a site adjacent to the University or, alternatively, for the Medical School to be moved from the University site to the hospital. The conflict concerning his proposal endured for many years and during his time in Melbourne Berry failed to carry the debate, despite much support from the medical profession.

One of the main instigators of these clinical schools of medicine was the Rockefeller Foundation, starting with the foundation of the Rockefeller Institute for Medical Research in 1904 in New York, a privately funded research institute with clinical facilities. In Britain the Wellcome Institute was established in the 1920s along similar lines and it is now among the largest supporters of medical research worldwide. In 1925 and 1926, Berry and (later Sir) Stanley Argyle, graduate of the Medical School, Chief Secretary and then Premier of Victoria, were funded by the Rockefeller Foundation to visit clinical schools in the United States.[81] Berry's report failed to win over the Council who were influenced by Barrett, despite the promise by the Rockefeller Foundation grant of $US250,000 to support the building of a new hospital. These funds then went to Sydney. Berry never saw his final victory, as the hospital move to the pig market site it occupies today occurred in 1935. B T Zwar, surgeon, part-time lecturer in Anatomy and Berry's ally in the pig market plan for the hospital, wrote to Berry on the 5 January 1945. He described the events that had resulted in the building of the Melbourne Hospital (later the Royal Melbourne) on Berry's preferred site as 'an achievement in which you played a memorable and noted part.'[82] Barrett attempted to rewrite history in his Halford Oration on 10 December 1932, claiming he had been behind the move.[83]

Berry's teaching and collecting

Among the innovations and improvements Richard Berry brought to Melbourne were the avid collection of anatomical and anthropological specimens, and the use of innovative techniques in the delivery of educational information. The original techniques he introduced were slides with texts, phonographs and moving films. The students commented in May 1908 in 'The Phonograph':

He is wound up tight, and going
On the muscles, nerves and bones—
Not the least attention showing, To our curses and our groans;
And we listen at dissection as he tells us all our sins,
Till we feel cold shivers creeping from our backbones to our shins.
O, he's stimulating—very.
Who? Prof. Berry.

Just before the lecture starts, he
Slips a record on his wheel,

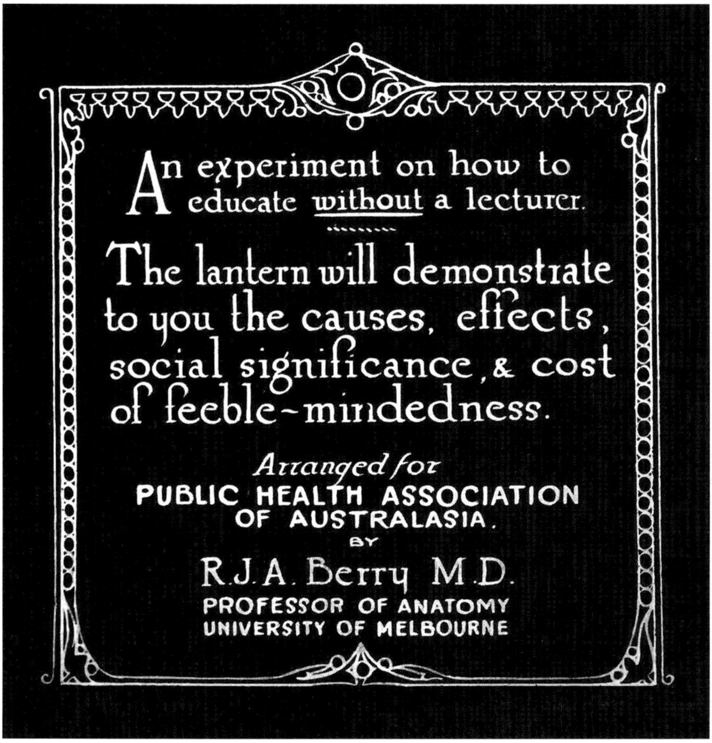

A lantern slide with text, one of Berry's innovative teaching aids.
Courtesy: Royal Australasian College of Surgeons Archives.

Then with mighty names of parts he
Makes our seven senses reel;
For he says—and we believe him—that he's like a phonograph,
And he seems to think it funny, but it doesn't make us laugh,
Wishing him in—Pondicherry,
Who? Prof. Berry.

When he sweeps aside conventions,
Put it down to the machine;
For the subjects that he mentions,
On the cylinder have been;
Yet we fancy at exam time,

> When we're on our beastly own,
> That we'd rather like a record from his giddy gramophone.
> Still, here's a long life and merry,
> To you, Prof. Berry.[84]

Berry's principal use of new technology was presented in his public lectures. He must have been a very early proponent of the cinema, and his film on mental deficiency was shown in the Melbourne Town Hall about the time of the First World War. He also produced sets of slides with printed text and pictures on eugenics and other topics. This, according to his introduction to a meeting of the Science Club at the University on 3 April 1925, was a method of 'lecturing without a lecturer'. According to *Farrago*, in such a lecture:

> Explanatory slides are exhibited, the length of the time they remain on the screen being proportional to the number of words they contain, and these are followed by pictorial slides in illustration of the points they made ... In proof of this contention, he [Berry] left for lunch before the lecture proper had commenced.[85]

Berry also wrote a textbook of Anatomy that was used by the Melbourne students for many years.[86] Today, in the Allen Museum of Pathology and Anatomy, Berry's wonderful sagittal sections of bodies prepared by his assistant, William Preston, are on display. They constitute the most impressive of many additions he made to educational material in the Medical School.

Berry left the University with some bitterness as he was neither offered the normal extension of five years at age sixty, nor was he made emeritus professor until his nineties, which he termed 'an almost posthumous honour.'[87] Berry was an exceptional teacher and an active and influential voice for the University in the community. He established the Anatomy Department as an effective and influential institution. Although he made many enemies, Berry deserved better treatment than he received from the University. While Barrett might have engineered his departure, no individual could have managed that. William Osborne became the Dean of the Medical School on Berry's retirement. *Speculum*, reporting on Berry's farewell dinner with the students, pointed out that a 'noted

Anatomy School workshop for preparation of specimens (note the bandsaw used for Berry's sagittal sections). Courtesy: Medical History Museum, The University of Melbourne.

absentee was Professor Osborne.'[88] Osborne was not invited to any other of the farewell dinners either.[89]

Wood Jones

It has been established that Frederic Wood Jones did not care for the legacy of Berry, but in terms of the curriculum or structure of the Department, he changed very little. It also needs to be remembered that Wood Jones was Professor for seven years and Berry for twenty-three. Also Wood Jones was not interested in administration, committees or buildings. For example, during the mid 1930s the pressures of numbers became so great that he was unable to teach the dental students. This

Berry as a monkey on Osborne's back. Cartoon in *Speculum*, August 1928, p.77.

prompted the outspoken Professor of Dentistry, (later Sir) Arthur Barton Pilgrim Amies, to complain to Vice-Chancellor Priestley:

> Amies came in with a diatribe against W[ood] J[ones], chiefly on his treatment of the dental students, but he said a good deal on general grounds also. I told him I did not share his opinion, but it has to be admitted that on the administrative side W[ood] J[ones] leaves something to be desired.[90]

Wood Jones wanted an environment that offered him the opportunity to conduct his research, speak in public and publish his work. During his time at Melbourne University he published over eighty articles. As such, he was not that involved in delivering the nuts and bolts of the anatomy course, leaving it to others to deliver most of the lectures and supervise the dissecting room. Significantly, Anatomy continued to be the pre-eminent subject in the pre-clinical course. In his memoirs of the 1930s as a student in the Medical School, David Jackson wrote:

> The life of the pre-clinical student in those days centred on the Anatomy School where we spent most of our time, dissecting. Every morning I would walk (usually with Doug Leslie my dissecting partner) from Trinity, across Tin Alley, by the University lake, under the fig trees outside the old Union Building and then by a path across a grassy open space to the Anatomy School. ... One knew the others in one's year to varying degrees, mostly the others who worked on the same cadaver (I think we were twelve on a cadaver.)[91]

Wood Jones preferred to work with senior and talented students, such as the prosectors. His weekly lecture, the 'weekly treat', was delivered on topics that interested him, hence the big turnout from interested outsiders.[92] Most students perceived him as a distant and elusive figurehead. As Jackson wrote of his experience in the dissecting room, '[t]he professor (Wood Jones) whom we all revered would sometimes look down on us from a high window.'[93]

Wood Jones also had little influence in improving the staffing problems inherited from Berry. Although student numbers had dropped by 1930 to nearly half those during Berry's tenure, the downwards trend reversed and by the mid to late 1930s large numbers of students were again enrolling in medicine. When Wood Jones arrived one hundred and seventy students were dissecting. By 1931 these numbers had grown to two hundred, and he reported to Arthur Keith:

> It is a queer thing that now I am getting on for 53, I am working far harder than I did at 35. I start in my department every morning between eight and half past and leave most nights a week after six. I have over 200 students dissecting and attending lectures and Hughes Jones is with you. Not that I regret it: but I am having a tough year, for I have to take all the Primary Fellowships classes myself and after three years in Honolulu away from topographical anatomy, that means a good deal of work. It is 15 years since I took Primary classes last at 8 Hunter Street, and it is quite an effort to go back in memory to the details one knew so well then.[94]

Near the end of his time in Melbourne, he found the pressure depressing. In correspondence to Keith he wrote, 'I have two more years of routine teaching and the deadly grind of having more than 300 students – scarcity of cadavers, an insufficient and underpaid staff, and an unsympathetic University administration.'[95]

When Wood Jones arrived in 1930 his staff included Hurley as senior lecturer in Histology; as Stewart Lecturer, the surgeon (and later chairman of the Royal Melbourne Hospital Board of Management) Eric Hughes-Jones; the surgeon Henry Searby, acting part-time as lecturer in regional and applied anatomy; part-time demonstrator in radiological anatomy, Hugo Flecker; and two surgeons acting as part-time demonstrators. He was no better staffed than during Berry's time. Flecker had been appointed to the Department when he offered an X-ray unit to the University in 1926.[96] Wood Jones admired Hughes-Jones and Flecker, and recommended them for work and publication in the United Kingdom, although of Flecker he wrote:

> He has ability. He has some racial and personal peculiarities that are hardly assets. He is sending you a long paper – its merits are beyond question: but is very long: its illustrations to my mind are futile and costly ... I say this only because the work comes from my department – not (and far from it) because I think it is not good work. It *is* good work. But Flecker is Flecker as well as being a disciple of Berry's.[97]

Flecker conducted pioneering research on highly poisonous jellyfish. He first named and worked on the 'Irukandji syndrome' and the previously undiscovered genus and species of lethal box jellyfish *Chironex fleckeri* was named after him.[98]

In 1935 when Wood Jones complained to the Council about staffing, Sydney Sunderland had replaced Hughes-Jones as senior lecturer, and the Stewart lecturer was Norman Harry. Sunderland was to follow Wood Jones as Chair in 1940, and before this appointment, the Acting Professor in 1938 and 1939 was Eustace Alwynne Rowlands. In 1938 and with student complaints about overcrowding ringing in their ears, Wood Jones, Ford and Sunderland all left the Department (Sunderland to study in Oxford) creating an enormous hole.

Royal Australasian College of Surgeons

It was no accident that Wood Jones was invited to present the inaugural lecture on the history of surgery at the opening of their first permanent headquarters of the Royal Australasian College of Surgeons in 1935. He had a particular relationship with Melbourne's surgeons that went beyond that of merely training them in anatomy. And he had been brought to Melbourne through the efforts of Alan Newton, a senior member of the profession. As the College had been recently established, it was important that Wood Jones trained those attempting to enter the ranks of surgeons and ensured their beginning was auspicious. As such, he was deeply involved in the early years of the College as examiner and lecturer, and his portrait by W B McInnes currently hangs in the foyer. He was also keen to establish high standards for surgery in Australia. This led him to publicly criticise the state of surgery before the establishment of the College in 1927.[99]

The radical advances in surgery in the late nineteenth century meant that operating became easier when patients were anaesthetised, and much safer after the introduction of antisepsis and asepsis. This meant that there was less need for the detailed knowledge of anatomy required by earlier surgeons, who had worked at high speed to minimise the effects of shock. Thus many general practitioners with little knowledge of anatomy practised part-time surgery and had often performed major operations. Many general practitioners would also refer their patients to unqualified practitioners for surgery and receive a portion of the practitioner's fee, although this practice was illegal. Therefore the attempt to introduce a specialist college to examine and accredit qualified surgeons was opposed vigorously by the most powerful doctors' union, the British Medical Association (BMA). At the BMA conference in Brisbane in 1920, setting up a college of surgeons was described as 'a dagger in the heart of the BMA'.[100] Under this cloud Wood Jones threw himself into the task of establishing the College as a successful body. The first examiners arrived in Australia to co-examine with Wood Jones in 1931,[101] and understandably he was nervous about his students' performance. He wrote to Arthur Keith, 'I do want my fellows to put up a decent performance ... in August.'[102] Without doubt, his efforts were a major contribution to the success of the College.

William Preston

It is impossible to discuss the teaching of anatomy in Melbourne without mentioning William Preston. I have pointed out that in this period the only significant and ongoing, full-time member of the academic staff was the Professor. William Preston, administrative assistant to both Richard Berry and Frederic Wood Jones, had been employed within the Department for the longest period of time. He had arrived in Melbourne around 1887 and became a surgeon's dresser (assistant) at the Melbourne Hospital, then dresser to Melbourne's first native born medical man to be knighted, Thomas Fitzgerald. In 1893 he became Harry Allen's laboratory assistant in Pathology, subsequently moving to Richard Berry's new Department in 1906. Greatly admired by the students and staff, he became the face of the Department because he filled so many roles. His skills in photography and other techniques of tissue preparation made him an invaluable assistant to Berry. Perhaps his greatest legacy was preparation of the transverse sections of the body under Berry's supervision, and these are still on view in the Harry Brookes Allen Museum today. He was well-known for his love of whisky and campanology, having been a bell-ringer at St Paul's Cathedral for decades. He was described in *Speculum*:

Frederic Wood Jones, Edward Ford, Sydney Sunderland, 1936.
Courtesy: La Trobe picture collection, State Library of Victoria.

William Preston invigilating at an examination.
Courtesy: James Guest.

> Our chief impression was one of waving arms and legs, surmounted by a fierce moustache that gripped a huge cheery wood pipe, but if a further mark of identification was necessary, it was furnished by a small faded cap that adhered (more than anything else) to his synciput [feature of the skull].[103]

His funeral in 1941 was a great affair with a large academic cortège, and muffled bells rang out as the procession passed St Paul's on its way to the Melbourne General Cemetery. There are many stories to tell about Preston. Sydney Sunderland recalls that he was a mine of whispered information to those favoured past students he accompanied to Masters practical examinations in surgery. Basil Kilvington became aware of the assistance Preston offered to some examinees, and having warned his fellow examiner, proceeded to present a purely fictional surgical procedure to the candidate. Preston, according to Kilvington, wandered over to the body, stared down at it for a minute and quite audibly said, 'Blimey doctor, I've never heard of it, you're on your own.'[104]

Wood Jones's legacy

A significant legacy of Wood Jones was the support and encouragement he gave to a number of young men in which, he believed, special qualities resided. He wrote to Arthur Keith in 1931:

> Fifty-two and working harder at teaching and helping younger men than ever I did before. But I love it all. I love it all, so long as there are young men coming along. That is where Honolulu failed for me. There was no chance of getting a good youngster and seeing him along ... So long as we can see a new generation coming and can help to shape them and guide them and see their strength – that must take the place (for me) of having babies of my own ... No man who does not beget working heirs (disciples if you will) loyal to him as children to a father, critical as modern children are apt to be with all their loyalty ... can ever be greatly considered in the history of any living sphere of activity.[105]

Among those he recommended to the scientific establishment in Europe were Hughes-Jones, Flecker, Edward King (later Professor of Pathology at Melbourne), Albert Coates and Rupert Willis (later a distinguished pathologist). It was Wood Jones who initially provided them with the entrée into the scientific world in which they were to flourish.[106] Perhaps his favorite was the brilliant anthropologist, Donald Thomson. Wood Jones wrote to Keith looking for assistance to publish his monographs on the Aboriginals of Northern Australia:

> Young Donald Thomson (if ever there was a case of racial heredity here it is) is completing a vast piece of work on the Cape York natives. The publication will be a problem ... I feel so much that this is first-hand and definitive work that I will do all I can to raise funds to assist its publication. Donald Thomson is a remarkable young man ... he would walk from Aberdeen to Cape York (probably having a row with someone every 20 miles or so) and arrive with no seat in his pants and no boots – *but he would get the minor things of the native culture:* and he has got them. Queer, loveable, heartbreaking fellow: I would hold up a rich squatter and demand his money to get that boy's stuff put out in decent form.[107]

Thomson was a former science graduate from Sydney and had a number of expeditions under his belt before he found a nurturing home in Wood

Jones's Anatomy Department, and a £500 per annum research fellowship that enabled him to return to northern Australia to continue his work. The flood of anthropological material for museums that resulted from this trip and a steady stream of research papers, coupled with Wood Jones's support, guaranteed a successful career and an exalted place in the history of anthropology in Australia. Thomson eventually became the first Professor of Anthropology at Melbourne in 1962, but the true significance of his work would take time before it was realised.[108]

A forum in which young talent could be recognised and nurtured was the McCoy Society. Founded in 1935 by Wood Jones, Priestley and others, the aim was to map the ecology of islands on field trips during vacations to Wood Jones's favorite destinations. A group of academics, postgraduate students and a few selected medical students would spend each day collecting specimens and mapping the geology, fauna and flora of the selected island. At night they would come together in their campsite and synthesise their findings through recording and discussion. The first year involved an expedition to Lady Julia Percy Island and the second to the Joseph Bank's group lasted for three weeks in 1936. The Royal Society of Victoria published both expeditions in reports.[109] Wood Jones believed that the Society was one of his best achievements in Australia, but concealed his University association because 'so long as I do not obtrude that side of it, I can get money to take good students to camp and to work.'[110]

Such associations were important for Wood Jones as he was largely unimpressed with Australia, but hopeful for the talent that he saw in some of his students. He told Priestley before he left the country:

> after 15 years in Australia he had accomplished nothing whatsoever. His outburst against 'the cross between pigs and policemen that are Australian politicians' was epic in its style … he has gained a reputation among the people that really matter in Melbourne, intellectually speaking, that is unique.[111]

In the 1920s his opinion was:

> Australia today is in a very strange case, it is living upon borrowed money and false ideals: it is living in an economic fools' paradise – not that these

> things matter to an anatomist nor influence his movements – but the unskilled man is everywhere and always the master of the trained and efficient, and the political power of the country is maintaining this state of affairs by every means within its reach.[112]

Conclusion

The three decades covered in this chapter in many respects represent the golden age of Anatomy in Victoria. In Richard Berry and Frederic Wood Jones, the University had anatomists of international reputation who drew the gaze of many outside Melbourne society to the University. They were men of great energy and imagination. Berry gained national importance through his involvement with the eugenics movement, as anatomical study was the keystone upon which eugenic theory was mostly based. He was one of Australia's first scientific experts in the field of public policy, and also put the teaching of Anatomy on a sound basis, thus cementing its central place in medical education in the state. Berry was also one of the first significant researchers in the Medical School.

Wood Jones continued Berry's vigorous approach. He advanced the cause of surgery in Australia, and even surpassed Berry's output of original work in a climate that was not conducive to research. Many of his Melbourne students, nurtured and supported by him, became significant international experts in their chosen fields. While lacking Berry's administrative skills, he continued the good work of the Department and brought Melbourne into the international circle at the highest level of anthropological and ecological research. He was also an early voice raising public consciousness of ecological degradation and the scandalous state of the first inhabitants of the continent.

The University of Melbourne, bereft of funds, found it difficult to adequately support the efforts of Berry and Wood Jones and ex-students filled the void. Much of the work carried out in the Department in this period is not recorded in this history, as few records remain. The legacy of those who undertook the work for little or no recompense is the survival of a Department that operated with few full-time staff and large student numbers. I made an earlier claim that this period was the golden age of Anatomy. A more accurate description, perhaps, would be the golden age of gross anatomy—the Anatomy of what the eye could see.

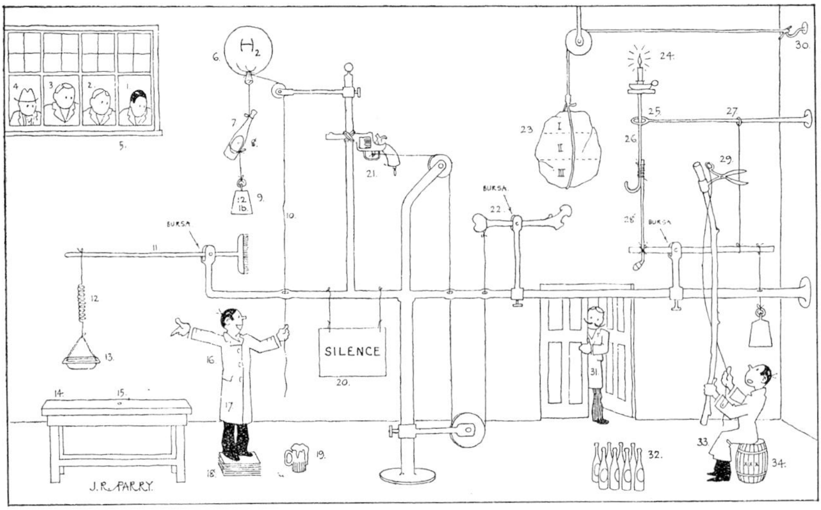

APPARATUS VIBRANS DISSECTIONIS

(O.T. Mechanical Rattling Machine for Students)

1. Professor iratus (sed benevolens).
2. Lector senior tolerans.
3. Lector Stewarti Fordi automobili vibrans.
4. Prestoneus jocosus.
5. Fenestra quadrata.
6. Vesica supportans.
7. Ampulla vitrea (Fosteri).
8. Inscriptio ovalis.
9. Moles pendens (atque gravis).
10. Ramus communicans.
11. Brumum Kennedii.
12. Trichina spiralis.
13. Dishum palpitans.
14. Mensa cadaveris.
15. Umbilicus mensae.
16. Studiosus regulans.
17. Toga alba.
18. Accumulatio Speculorum.
19. Amicus studiosi—semper causa jubilationis.
20. Exhortatio.
21. Revolva revolvans.
22. Magnum Bonum.
23. Petrus Galliae (O.T. Gaul Stone—divisa in partes tres).
24. Lux vomica.
25. Foramen ovale.
26. Scipio ambulans.
27. Foramen rotundum.
28. Scipio golfi.
29. Forfex domestica.
30. Hamatum murale.
31. Fallopius antiquus. (Dies Irae).
32. Ampullae reservatae.
33. Studiosus operans.
34. Niner rotundum (Nortoni).

Anatomy Department as seen by John Parry (medical student).
Cartoon from *Speculum*, May 1933, p.175.

Wood Jones knew that gross anatomy as a research topic was exhausted. In 1923 in an address at Adelaide University, he told his audience that the community of scholars at a university, rather than respecting the Professor of Anatomy, expected less of this individual than anyone else:

> So long as he is acquainted with the structure of the human body, as a cabman knows the streets of the city in which he plies for hire … but little more is asked of him … When we rank it beside the vast subject of physics,

> the endless possibilities of chemistry, or the modern developments of physiology, what a poor, limited, exhausted field it seems.

He then went on to argue, despite his negative proclamation, that Anatomy 'is the father of the physiologist, of the embryologist, the histologist and the anthropologist, among others.'[113] It was in these areas that he saw the anatomist extending the bounds of knowledge.[114] The period following Wood Jones's departure was to see the study of anatomy in Melbourne follow those paths, both in research and teaching, despite the passing of the golden age of gross anatomy.

CHAPTER EIGHT

A thoroughly modern professor, Sydney Sunderland: 'the confrontation of medicine as a science and medicine as an art'

On 25 July 1938, Sydney Sunderland wrote from Oxford to the Registrar of the University of Melbourne to accept the offer of the Chair of Anatomy, vacated in 1937 by his teacher and mentor, Frederic Wood Jones. Sunderland, a Queenslander, had completed a brilliant degree at Melbourne, topping every year and winning many prizes. He had been appointed a Senior Lecturer in Anatomy on graduation. The eminent surgeon, Sir Alan Newton, described Sunderland as, 'one of the most brilliant medical graduates of the University of Melbourne.'[1] Sunderland had also passed the Primary Fellowship Examination of The Royal College of Surgeons of England before completing his medical course. Born on 31 December 1910, Sunderland was only twenty-seven when he was offered the Chair.[2] The date of his expected return to Melbourne to take up his position had been delayed for a year. As the Melbourne press put it, 'so valuable is the work being done by Dr Sunderland at present that the Council is making the appointment effective from 1940 and not from March 1939, as intended'.[3]

It is difficult to ascertain whose idea it was for Sunderland to remain in the northern hemisphere. He had only begun his research and teaching in the Department of Anatomy at Oxford University under Professor

Wilfrid Le Gros Clark in 1938. The balance of probabilities, however, seems to indicate that it was Sunderland who wished to remain in England. The University Council had laid down conditions for his absence that required him to 'obtain further experience' and it was 'especially desirous that this experience shall include visits to medical schools on the Continent.'[4] To further promote the theory that it was Sunderland who was keen to stay in Europe, the Council initially refused to finance his stay. After a number of rejections Sunderland pleaded with the Council for money. He wrote 'a comprehensive survey would obviously be impossible if I were compelled to remain in Oxford continuously during 1939 because of economic reasons.'[5] Despite his many requests for funds, they eventually, and reluctantly, only allowed him an advance on his salary, later exacting interest on a portion of the repayment. Sunderland's financial relationship with the University was to continue to oscillate between extremes throughout his career.[6]

Whatever the initial problems were, however, one would expect that he would have been delighted with the University's offer of the Chair, engineered principally by Wood Jones, who had earlier put him forward for the Chair in Anatomy in Hong Kong in 1936.[7] Wood Jones wrote to Arthur Keith in that year:

> My boy Sunderland knows his anatomy in his way – he has had his nose rubbed in it. But in addition he knows his literature in most European languages and is particularly well up to date in clinical neurology. I hope he gets Hong Kong for he will go far. I should like him to take my job here when I give [it] up.[8]

Part of Sunderland's pleasure in receiving the offer would have been due to the position he found himself when in Oxford. Le Gros Clark seems to have become somewhat irritated with Sunderland. The reasons are not known, but it may have been about the time Sunderland was devoting to courting his fiancé, Gwen Johnston, rather than focusing on his work.[9] Sunderland may have been concerned that this could have potentially undermined possibilities for promotion in England. Another reason for his enthusiasm to visit a great variety of departments in the northern hemisphere was his desire to put his own stamp on the Department. This was something Wood Jones had never displayed any great enthusiasm for, preferring instead to promote individuals. In his letter of acceptance of the Chair on 25 July 1938, Sunderland wrote that he was:

> particularly anxious to develop the Melbourne school, not only as a teaching centre for post graduate instruction, but also as a first class research centre. In order to do this it is essential that while here I should undertake a complete and thorough investigation of all the modern methods of research at present employed in Continental and American centres.[10]

If it seems that Sunderland's views represent a rejection of Wood Jones's aversion to the laboratory, there is certainly no evidence to suggest this was the case. They continued to correspond, and Wood Jones unsuccessfully tried to have Sunderland made a Fellow of the Royal Society a decade after his appointment.[11] It seems there was some opposition to his admittance, perhaps from Le Gros Clark. Wood Jones also believed Sunderland spent too much time on administration, thus denying himself full research expression—research which Wood Jones continued to promote in England until his death.

Sunderland's time in Oxford, however, was not wasted. He was directed and encouraged in his future research career by the Nuffield Professor of Surgery and pioneer of neurosurgery, Sir Hugh Cairns, who was an Adelaide graduate. He also began a friendship with Pio del Rio-Hortega, the Spanish political refugee who introduced him to the most up-to-date staining methods.[12]

Sunderland's tour of 1939 and 1940 as the newly appointed Professor of Anatomy set the perimeters for his future career. He became an indomitable traveller, though not in Wood Jones's style, and during his visits to many leading laboratories, he made important and lasting friendships. Many were stars in the fast developing neurological world, such as Wilder Penfield in Montreal; those at the neurology laboratory at Harvard; the neurophysiologist, John Fulton, at Yale; and the neurosurgeon, Earl Walker, at Johns Hopkins.[13] His extended stay in the United States also created an ongoing relationship with that country, cementing it as a future favoured destination, rather than the United Kingdom and Europe, thus trailblazing a new trend in Australian intellectual life. Throughout his career Sunderland, and almost invariably his wife Gwen, regularly travelled the world on the new domestic and international airlines that proliferated after the Second World War, attending conferences and visiting fellow neurologists, neuroanatomists and neurosurgeons. Their charming and easy manner made them welcome guests, attested to by the innumerable enthusiastic correspondence from those encountered. For example, Louis B Flexner, world leader in the study of memory and learning, wrote to Sunderland on 21 January 1954, 'Should the clouds clear and the sun shine through do let me know because several of us would be delighted to see you again. When you are about everyone that sees you is stimulated by your enthusiasm.'[14]

Home

The period following his return to Melbourne in early 1940 was to become important in a number of ways. Due to reductions in staffing and resources caused by war, he was to learn the hard way the importance of gaining political influence, both within the University and in government circles, so as to place his Department (and later Faculty) in the best possible position. Most importantly, this period saw him develop an interest in peripheral nerve injuries due to the large number of cases he examined among repatriated soldiers. The study of peripheral nerve injuries became his major lifelong research project, bringing Sunderland and the Medical School international recognition, and later fame. The clinical work also inspired in him an understanding of the importance of the non-scientific side of medicine. As testimony to this, he continued this work with patients for significant periods throughout his career. Although he was a leading researcher when scientific medicine reached an apogee, he was, however, acutely aware of the dangers involved. He told the first graduates of the University of Tasmania Medical School in 1970:

> in the practice of medicine the doctor should aim at more than the cultivation of an intellectual excellence. Much that is important is bound up with other highly valued human qualities which relate to the management of the patient as opposed to the management of his illness. In essence this is the confrontation of medicine as a science and medicine as an art. This issue is particularly important at the present time when the increasing emphasis on laboratory procedures and the scientific aspects of medicine is shifting the centre of interest from the patient to the illness.[15]

Professor, teacher, researcher, 1940–53

Sunderland was an impressive, hard-working teacher. In 1946 he received a petition from a large group of students requesting that he reinstate a course on Neuroanatomy 'because of its great reputation.'[16] Others remember the great care he attached to his students' wellbeing. When Keith Henderson (later senior neurosurgeon at St Vincent's Hospital), spent six months in bed with peritonitis and missed the Histology course, Sunderland organised and supervised a special course to fill the gap.[17] These qualities had been obvious from his years as a student when he first started teaching in the Anatomy Department. In a letter written by the surgeon, Sir Hugh B Devine, for Sunderland's application for professorship at the University of Hong Kong in 1936, he remarked:

> His demonstrations and lectures have been a feature in the teaching of Anatomy in the Melbourne University. He has a pleasing and dignified personality, and has shewn [sic] that he has a good influence with students and that he can control them; and, from my personal knowledge, I know that he commands their respect and admiration.[18]

As far as teaching in the Anatomy Department upon his return in 1940 was concerned, Sunderland was up against it. On 21 June 1952, he described the conditions in a letter to George Paton, the recently appointed Vice-Chancellor.[19] Sunderland had requested a grant of £650 to travel abroad from the University Council controlled Special Anatomy Fund. This had been rejected on the basis that departmental staff had raised the money, and so it should not be spent on any individual.[20] Sunderland was furious and wrote at length to right the wrong he felt had been attributed to him. At the time Paton had only been Vice-Chancellor for one year, so Sunderland pointed out that when he had arrived at the Anatomy Department in 1940:

> Within the first two terms I had lost every member of the department with any training in anatomy, embryology and/or histology. Over this short period there were also 10 staff changes occasioned by demonstrators joining the staff for short periods of time pending their call-up. Thus I was left to conduct the affairs of a large and important department without the

> assistance of experienced personnel and with rapidly changing staff which dislocated teaching and department administration. Furthermore, I had no permanent assistant … it has only been possible to build up a full-time staff in the postwar period.[21]

Sunderland further pointed out that 'as the only trained anatomist in the department my duties and responsibilities greatly exceed those which could reasonably be expected of the Head of a department even under an emergency … no other department in the University suffered … to nearly the same extent as mine'.[22] Not only was he overworked, but he had also saved the University thousands of pounds in salaries. As if this was not enough, Sunderland was also teaching the massage (physiotherapy) students during the war and later taught speech therapy and optometry. A special government grant was paid to the University for this teaching, but Sunderland had forgone his payment as a voluntary war measure and established the Anatomy Fund into which his money had been placed. All other teachers, however, were paid. Sunderland concluded:

> £650 is a large amount to request from a fund of £725 but I believe it is entirely justified by the circumstances. By my efforts during the war years the Department continued to function under exceptionally trying conditions which, however, meant very considerable savings in salaries for the University … Finally, the sum of £650 was specified because this is the amount required to cover my return airfare from London via America.[23]

Sunderland received his money, but he had struck a bad case of institutional amnesia due to the retirement a year earlier of the charismatic John Medley. Sunderland may have felt that his early sacrifices had been forgotten and that the particular debt owed by Council would be ignored. Hence he wrote to Paton, 'I hope that this rather lengthy, but necessary account of the background to my request will explain why I felt justified in making the application.'[24]

In fact the debt owed to Sunderland for his initial labours as professor had been recognised as early as 1941, when Medley wrote to Sunderland:

> I appreciate what you have done in your Department since you took charge of it, and I know that my appreciation is shared by the members of the

> Council. I can only hope that in the future it will be possible to do more for you than it has been in the past.[25]

If there had been any animosity between Sunderland and Paton in 1952, it was resolved quickly. In late 1953 Paton told Sunderland 'the Council is greatly indebted to all members of the staff who have put up with much inconvenience in recent years through the unsatisfactory buildings, large numbers of students and low salaries.'[26] This respect for Sunderland lasted the full period of Paton's long reign. In 1965, and during a period of illness, Sunderland received a letter from Paton wishing him a full recovery and encouraging him to stay at home for as long as was deemed necessary because 'you are too valuable!!'[27] The support of the Vice-Chancellor was important as Sunderland would plan and execute great changes to the Medical School during his tenure as Dean. As Poynter and Rasmussen have stated of the period dominated by Sunderland's reign:

> The [Medical] Faculty's Dean came to speak and act in a manner rare among other deans, whose role too often resembled that of a presiding officer of a debating club ... [the Dean of Medicine] made colleagues who questioned ... [the Faculty's] wisdom look like players in a lower league.[28]

An intriguing aside in Sunderland's letter to Paton was in regard to current staffing in 1952. He told Paton, apropos of the unsatisfactory arrangements concerning his workload during the war, 'now that I have built up a full-time and capable staff, I hope to retire from this section of departmental responsibility.'[29] Returning to reasonable staffing levels had obviously taken many years.

After the war the Department returned to staffing levels similar to those under Wood Jones. Staff numbers however, did not rise in relation to the increasing number of students enrolled. As after the First World War, Medicine proved popular with returned service members. In 1945 there were six hundred and seven medical students, and by 1946 this number had increased to seven hundred and forty-eight; in 1947 to one thousand and eight; and reached a peak of one thousand one hundred and twenty-nine students in 1951, significantly more than at any other time.[30] The enrolments in the 1950s and 60s were affected by important policy changes to be discussed later in this chapter.

The full-time members of the staff in 1946 (other than Sunderland) were Kenneth F Russell (senior lecturer); J A Larwill (senior lecturer in Histology and Embryology); Leslie J Ray and A G Matheson (lecturers); and Gordon Keys Smith (demonstrator/ lecturer) who left to practise as a missionary doctor in South East Asia in 1949, later returning to the Royal Children's Hospital as a surgeon. Matheson left in 1947 and Ray was made a senior lecturer in 1948. Part-time staff included R S Hooper (lecturer until 1947), Albert E Coates (lecturer in Surgical Anatomy), and the polymath John O'Sullivan (demonstrator in Radiological Anatomy). By 1948 George Bedbrook (later Perth surgeon and pioneer of paraplegic sports) was a lecturer, J B Turner had replaced Coates and a number of demonstrators had been appointed. They were K C Bradley, W E Swaney, N C R Merrilies, W R Hayhow, E L G Beavis and A F Roche.

Perhaps the most distinguished member of staff was Arthur Schüller, who held an honorary position.[31] Schüller was a Viennese medical graduate and pioneering radiologist who fled Nazi persecution in the 1930s. He was unable to have his qualifications recognised in Australia, although he was a brilliant graduate from Vienna University, and had received the title *sub auspiciis imperatoris*, meaning a doctor under the patronage of the Emperor Franz Joseph, an award bestowed only twice in his sixty-seven-year reign.[32] Schüller has been named 'without any shadow of doubt, the father of neuroradiology'.[33] When he arrived as a refugee he was sixty-five years old and penniless. He worked in an honorary position in the Radiology Department working with, and advising, a number of staff, including ex-student and radiologist, John O'Sullivan and the neurosurgeon Frank Morgan. He was paid in such a way that no records ever appeared in any financial statement. It is not known to this day where the money came from, though it has been suggested that he may have been partly sponsored by those he advised. Schüller continued to produce research until his death in 1957.[34]

I have listed the staff in the Department immediately after the war because a number were to become key members throughout the decades of Sunderland's tenure as Professor and Dean. Among the demonstrators was Alex Roche who set up the University of Melbourne Child Growth Study in 1954. He became a Reader in 1962 and later moved to Wright State University, Ohio, in the United States, subsequently holding a number of Chairs in Community Health, Paediatrics, Obstetrics and Gynaecology.

Russell and Ray were to serve in the Department throughout Sunderland's era, and both attained professorships: Ray in Anatomy and Russell in Medical History. Both represented the older Wood Jones tradition, taught gross anatomy and continued the Department's interest in comparative physical anthropology. They were caretakers of the Departmental collections of human skeletal remains, although neither produced any significant research on the collections. A classmate of Sunderland, Russell's interests from early on in his career were antiquarian. As his colleague, Smith recalls 'although he was a good anatomist, anatomy was not his main interest; his main interest was medical history and this was evident even when I was on the staff.'[35] Apart from his antiquarian research, Russell wrote the *Centenary History of the Medical School* (1977) and was an important figure in the creation of the Medical Museum and the fine collection of antiquarian books in the Medical Library. Similarly, Ray assisted Sunderland in his research into peripheral nerves. His first interest, comparative anatomy, increasingly took second place to the pressures of teaching and administration. Also, his interests in anthropology were overshadowed by the Department's concentration on the study of the peripheral nervous system.[36] His friend, Smith, wrote of Ray:

> the dominant interest in the Department was the nervous system and he was thus sidetracked when he should have been developing his skills in the field of Comparative Anatomy, and, in later years, he never really recaptured his enthusiasm and expertise in that area.[37]

The weight of teaching and administration and the burdens imposed by the building program and subsequent expansion of student numbers in the 1960s further restricted his opportunity for research later in his career.

Keith Bradley, another member of staff appointed after the war, lectured to the dental students. At this time he was studying neurosurgery under Hugh Trumble at the Alfred Hospital, and replaced him on retirement. Bradley continued to conduct research on Aboriginal skeletal remains collected throughout the Department's history, as well as investigating the phenomenon of phantom limbs, a subject close to Sunderland's work. Bradley co-authored a number of publications with Sunderland, but continued to work mainly as a surgeon after becoming a professor in the Department in 1967.[38]

W Ross Adey taught in the Melbourne Department for almost a decade until he became Professor of Anatomy and Physiology at the University of California, Los Angeles in 1957, and a leading figure in research on the brain and electromagnetics.[39] Neil C R Merrilies began his training as a plastic surgeon at the Heidelberg Repatriation Hospital after the war, but then switched to Histology and taught this subject in the Department until his retirement in 1979. His major study focused on the extrinsic muscles of the eyeball. W R Hayhow worked in the Department briefly and then moved to teach Anatomy in the Medical School in Brisbane in 1950. John O Lavarack was another long-term member of the Department, moving from Pathology to Anatomy in 1949. His interests in the microscopic appearance of peripheral nerve fibres after injury, as well as their functional correlation, naturally complemented Sunderland's work. He was appointed Senior Lecturer in Histology and Embryology, then Reader in 1956, before his retirement in 1979.[40] Among the last appointments made by Sunderland was Geoffrey C T Kenny in 1959. An early Queensland graduate, Kenny began studying the pineal gland under Le Gros Clark in Oxford, and continued in this field until his retirement in 1990. Kenny was very active in the area of medical history and instrumental in the establishment of the Australian History of Medicine Society in 1986, of which he was later President (1993–5). He became Deputy Head of the Department in 1982 until his retirement in 1990. Others who worked in the Department briefly in the 1950s include O J Lewis who left to work in London, and C J Barnett who also departed for London, and then returned as the inaugural Professor of Anatomy at the Medical School in Hobart. Colin Dunlop worked with Adey for some time and left in 1962 to take up a position in the United States. Clearly Sunderland's efforts to create a research as well as a teaching Department had met with success.

Among those part-time staff after the war was James Guest who became a surgeon and was based at the Alfred, but spent hours every week in the Department for many years demonstrating, lecturing and conducting research. While William Swaney also became a surgeon, he continued the tradition of ex-students teaching anatomy, in his case demonstrating and lecturing on the skeleton, something he continued on a part-time basis throughout his career.

Research

Sunderland was a man with many strings to his bow. Although he faced great difficulties in the teaching and administration of the Department in the early years, he was determined to continue with active research, as he had promised the Council. While Wood Jones had been one of the great observational biologists of his age, continuing a line of morphology reinvigorated by T H Huxley, Sunderland incorporated into this observational comparative anatomy a new approach, following a path of experimentation and the application of modern techniques. This approach was perfected in his work on the peripheral nerves. He did, however, keep the Berry–Wood Jones physical anthropological stream of research alive in the Department, though he published virtually nothing on this topic himself.[41]

Sunderland began his research under Wood Jones, and in 1932 he presented a paper on the brain of the Madagascan chiromys (a mammal) to the Anatomical Society, founded by Wood Jones. This work was later submitted to the *Journal of Anatomy* in London and its rejection caused a considerable furore. At the same meeting, Kenneth Russell (later a professor in Sunderland's Department), presented a paper on the 'Rotation of the Eye', and *Speculum* noted 'A little lighter vein was introduced when Mr. Russell's tortoise got shy and would persist in withdrawing its head from the public gaze.'[42]

Sunderland's greatest period of research activity commenced on his return to Melbourne in 1940, and this continued unabated until his appointment as Dean of the Medical School in 1953, when his research slowed considerably due to administration. This impediment to his research was never really removed, as he took on even more duties relating to the university and the nation throughout the 1950s and 60s. He did not abandon his interest in research, however, nor did he cease publishing.

A brief and fascinating interlude in the late 1950s and early 60s found him studying Kuru, demonstrating his range of research interests, even when he was burdened with administration. Kuru was the mysterious brain disease discovered in remote parts of the New Guinea highlands by D C Gajdusek. Sunderland provided a neuroanatomical report for Gajdusek in 1958, and in return, he received a copy of Gajdusek's (then unpublished) New Guinea journal, inscribed 'to Sydney Sunderland in memory of your early visit to see Kuru.' This disease proved to be caused by a prion connected to cannibalism (similar to the type of organism attributed to 'Mad Cow' disease).[43]

Perhaps Sunderland's finest works, *Nerves and Nerve Injuries* (1968, 1978), and *Nerve Injuries and their Repair* (1991) were produced during this period of great administrative activity and into his retirement. As an academic in a field where research has a limited shelf-life, Darian-Smith noted that in the period 1991 to 1995, the average number of citations of Sunderland's papers was one hundred and ten per annum; a truly remarkable feat.[44] After the war, and due to the growth in new technologies, it became more common for research to be collaborative, with a professor leading a team of researchers rather than working in isolation. Many of Sunderland's papers were the result of such collaboration with Ray and Bradley. In that sense, Sunderland was following his own advice by not approaching problems in a limited fashion. Thus others in the Department deserve to share part of Sunderland's reputation in the area of peripheral nerves.

It was fortuitous for Sunderland that all Australian soldiers in 1940 suffering from chronic peripheral nerve injuries – a common affliction in wartime – were accommodated at the 115 Army General Hospital in the Melbourne suburb of Heidelberg.[45] By this time antibiotics had arrived, and this allowed the microscopic surgical research Sunderland would undertake. He accepted an offer to act as a clinician for the soldiers—a role he eventually cherished. He immediately began collecting case studies and observing the progress of his patients.[46] In an early undated lecture, he spoke of this five-year study on over three hundred peripheral nerve injuries,[47] outlining the competing theories of nerve repair. The first theory, that of Foerster and the French school, was that attempts at nerve repair should be conservative and nothing should be attempted in the first six months. The second theory was the Oxford method championed by Seddon, advocating early intervention.[48] Much of the surgery on soldiers in Melbourne had been undertaken by Sunderland's mentor Hugh Trumble.[49] Sunderland concluded that as about seventy per cent of damaged nerves had regrown after six months, conservative therapy provided a much better outcome than early intervention. Further he concluded: 'the chief concern of the surgeon should be to respect the continuity of the nerve and always bear in mind the delicacy of nerve

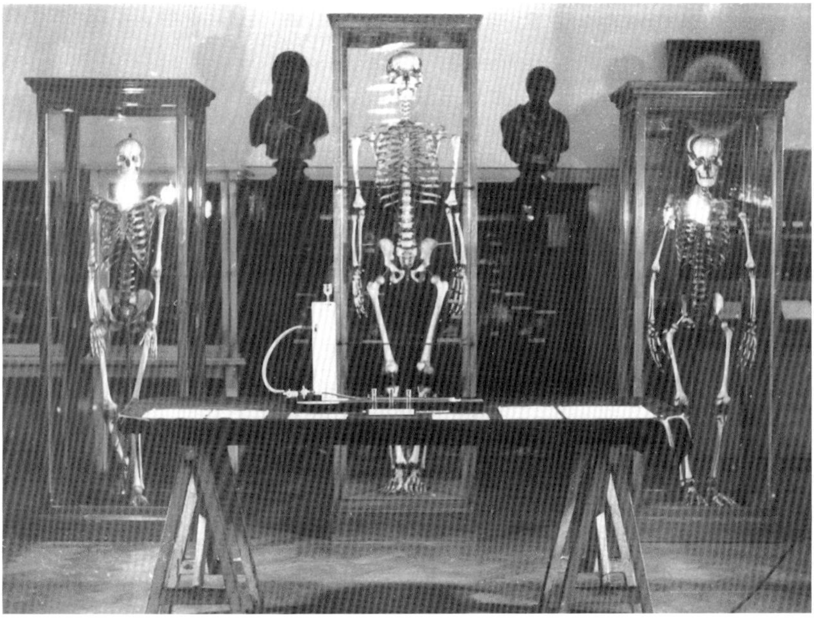

Anatomy Museum, *circa* 1950.
From the collection of the Department of Anatomy and Cell Biology.

tissue, the difficulty of its repair and the unsatisfactory end results of suture.'[50] He was to essentially maintain the same position, even after intensive study of peripheral nerves had been undertaken in the decades that followed.

Although he specialised in peripheral nerves, Darian-Smith points out that Sunderland: 'retained wide research interests ... [in] topographic anatomy, structure of the cerebral cortex, the connections of the hypothalamus, the vascular supply of various organs and tissues, the pupilloconstrictor pathways, and medical education.'[51] Medical education was Sunderland's interest in the second half of his career. Sunderland was a curious researcher after knowledge. He was always interested in and connected with the Melbourne Zoo, being the honorary anatomist and later a board member. This meant the promise of interesting animals to study, including an elephant's brain on one occasion. A young, fully grown lion with a brain haemorrhage resulting in partial paralysis was also sent to the Department. Smith, demonstrator at this time, wrote:

Professor Kenneth Russell inspecting museum exhibit. Russell developed the museum and Rare Book Collection. Courtesy: Medical History Museum, The University of Melbourne.

anthropology – finger prints & my old friends Talgai & Cohuna ... all fresh, welcome and precise additions. My thanks to you & congratulations to the coming men'. Finally in 1954 Keith wrote, 'I do congratulate you on the quantity & quality of the research that comes from you and your Department.'

Sunderland was also the subject of an attack concerning his qualifications in the early years. This setback involved the notorious 'pyjama girl' murder. It seems in retrospect that the investigation into the murder of Linda Agostini was riddled with tampered evidence, police dishonesty and corruption.[58] Among many problems encountered was the inconvenient fact that the preserved corpse of Linda Agostini, the so-called brown-eyed 'pyjama girl', had blue eyes. Sunderland was called in as an expert witness at the hearings in 1944 to explain away the problem. The ophthalmologist, Kevin O'Day, opposed Sunderland and legal counsel questioned Sunderland's expertise (as his only qualifications were a MB BS). Sunderland was determined to eradicate this weakness in his profile and gained his Doctorate of Science from published work the following year.[59] By 1961 he was recognised as a world authority in neurology, and was asked by the Wellcome Trust to provide a report on the state of research in the neurological sciences in research institutions in the United Kingdom. He first enunciated in his report what he believed was the best technique for solving neurological problems. He argued:

> Future progress in the biological field will depend in large measure on our ability and capacity to bring every available method to bear on any particular problem. In doing so there is no place for the rigid subdivision of biology into morphology, physiology and biochemistry. Problems know no such subdivisions, and restrictions, financial and otherwise, which limit those working in anatomy departments solely to the morphological aspects of a problem are wasteful, frustrating and sterile ... If progress is to be accelerated, particularly in neurophysiology, anatomists must be encouraged to extend their enquiries into the functional field, while physiologists should develop a healthier respect for structure than appears [to be] the case.[60]

When summing up the English scene Sunderland was not complimentary, as he observed that in 'most British schools a sharp line of demarcation still persists, unfortunately between structure and function whereas in

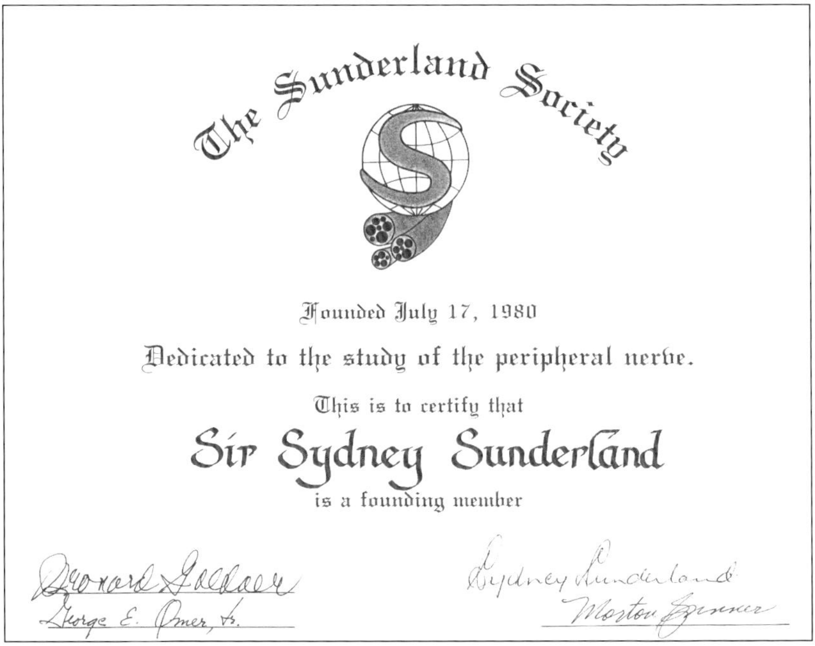
The Sunderland Society

Founded July 17, 1980

Dedicated to the study of the peripheral nerve.

This is to certify that

Sir Sydney Sunderland

is a founding member

George E. Omer, Jr.

Sydney Sunderland

Morton Spinner

The Sunderland Society (an international body) was formed in honour of Sunderland's seminal work on peripheral nerves. Courtesy: The University of Melbourne Archives.

> young people like Adey who are really going to do the most significant work in the future. There are a few people like Jack Eccles, of course, who work twenty-four hours a day, peering into a cathode ray oscillograph but they are not very numerous. You are so good at the other thing and your talents are so widely recognised, as witnessed by the series of appointments that have come to you, that I don't know how much time you are going to find to give to the laboratory even though you do give up the Deanship … It certainly isn't any fun to feel that one is falling behind in neurology.[66]

Clearly Sunderland had reached a defining moment in his career. The burdens of performing the administrative duties of both Chair in Anatomy and the Deanship were eroding his research time. Rather than retire, however, he continued with the Deanship and did not relinquish the Chair until 1961. In just under a decade, he also took on major responsibilities with the Australian Academy of Science, a major reform in medical education in Victoria, and a number of national responsibilities beginning with the foundation of a medical school in Bukittingi on the island of Sumatra in Indonesia.

Bukittingi

On 23 May 1956 Sunderland received a letter from Frank Fenner, Professor of Microbiology at the John Curtain School of Medical Research, Australian National University, Canberra.[67] Fenner had a number of connections with Sunderland and, in particular, he was a great admirer of Wood Jones, having been inspired by him as a student at the Medical School in Adelaide. Fenner had also met his future wife Bobbie at the Heidelberg Military Hospital (where Sunderland's peripheral nerve injury unit was located) while working at the Walter and Eliza Hall Institute of Medical Research under MacFarlane Burnet. In a highly distinguished career, Fenner's greatest contributions were his seminal work on myxomatosis and leading the World Health Organization team that eradicated smallpox in 1969. Fenner wrote to Sunderland to explain that he had received a letter on 11 May 1956 from the Department of External Affairs explaining:

> For some time we have been receiving from Indonesia enquiries [as to] whether we could assist under the Colombo Plan in staffing and equipping Indonesian universities ... We are anxious to help and our Ambassador in Indonesia ... is convinced of the wisdom of the policy of decentralising the medical and other science training institutions on the grounds that the final answer to Indonesia's great need for technically trained men is the establishment of training institutes within Indonesia itself.[68]

Much of the assistance administered under the Colombo Plan involved bringing South East Asian students to universities in Australia and other Western nations. At the Melbourne Medical School there were seventy-one Asian students enrolled in 1955.[69] This was, therefore, a radical suggestion. Fenner was asked to lead a delegation of three eminent medical administrators to assess the Indonesian need. He wrote to Sunderland asking if he would be a member of the delegation who would conduct the survey in the August vacation of 1956. Sunderland replied in

Bukittingi (Sumatra) Medical School under construction.
Courtesy: The University of Melbourne Archives.

the affirmative.[70] This would be the first of many projects that he would undertake for the Australian Government. The third member of the delegation would be H N Robson, first Professor of Medicine at the University of Adelaide. Fenner then provided Sunderland with a number of assessments on Indonesian education, including a report prepared by the Canadian Government. He warned Sunderland that this report should 'not come to the notice of the Indonesian Government as it is rather critical.'[71] The report identified Bukittingi as the preferred sight for directing future aid, and further detailed that a medical school had been opened in Bukittingi in 1955 that consisted of:

> a small school building which contains a few lecture rooms and two staffrooms ... in January 1956 there was no library and no equipment. Most lectures then were given in the local state high school ... Enrolments consisted of 60 first year students (15 of them girls). It has been very difficult to recruit staff. Lectures are given by local doctors and [state school] staff ...[also] The Medical Faculty in Bukittingi is much farther behind in its

> building programme than the Faculty of Agriculture. It was therefore surprising to hear so much about the lecturers and professors and the other aid from Australia which the Faculty expected at this early stage.[72]

The report recommended the development of Bukittingi because of the promise of land on which to build, good cultural amenities, an excellent climate, good high schools and 'the great interest in education shown by the inhabitants and the authorities in that town.'[73] Further, the Canadian report recommended that all Canada could do, 'due to the shortage of university staff', was to take students into its own universities. A major involvement in Indonesia, however, would 'be to the detriment of Canada's own universities'.[74] It noted that the University of California was already involved in projects in Indonesia.

Situated in Central Sumatra beneath an active volcano, 'Bukit tingi' means 'high hill'. The area in which it is situated in the Minangkabau Highlands is 'of great natural beauty',[75] about one thousand metres above sea level, and noticeably cooler than the steamy coastal plains of Sumatra. Traditionally a matriarchal society, the locals are proud and independent of spirit, interestingly providing 'a large proportion of the political and intellectual leaders of Indonesia'.[76] This trip was to be a seminal experience for Sunderland, as it confirmed his position as a staunch advocate of supporting and encouraging the development of medical education in Asia. In a lecture at the Nurses' Memorial Centre, St Kilda Road, on the evening of 26 July 1958 for the Australian Asian Association of Victoria, Sunderland:

> urged a greater knowledge of [Asia's] traditions and artistic attainments, sought a more intelligent appreciation of their problems and aspirations and expressed the view that much of the assistance we could offer would be better given on the spot rather than bringing students to Australia for training.[77]

On 6 September 1956 the Report on Medical Education in Indonesia by Fenner, Sunderland and Robson was presented to the Department of External Affairs, summarising their visit to Indonesia between 25 July and 17 August. The aim of this trip had been to conduct a quick survey of the six medical schools (including three previously established schools on

Java), and spend more extensive time examining the Bukittingi project.[78] They concurred that 'Academic standards are apparently being maintained at a high level'.[79] However, Fenner, Sunderland and Robson saw a great need for reform:

> One of the most obvious features of the University scene is the very large number of students, both in the established and in the embryonic schools. About 21,000 were enrolled in 1955. Admission to University is not controlled, and any student who graduates from High School ... is entitled to proceed to the University. [80]

Almost all of the students (ninety-nine per cent), however, failed to graduate, and this was partially due to the need for students to work long hours outside study to earn enough to live on. This failed to discourage the aspirants because, according to the Report, 'there is general evidence of a widespread and intense desire for education at all levels' most probably because 'very considerable weight is laid on academic qualifications in the community, and status and salary are almost wholly determined by the seniority of the degree held.'[81]

More specifically, their view on medical education was that the present institutions were entirely inadequate for the purpose of fulfilling Indonesia's need for medical care; having produced only one thousand five hundred medically qualified personnel (mostly based in the large cities) for a population of eighty million.[82] And it was almost impossible to recruit staff for medical schools as the pay was poor and at least ten times as much income could be made in private practice.[83] As far as pre-clinical teaching was concerned, it was 'to a very large extent dependent on foreigners working in Indonesia under various auspices', significantly financed by American money and personnel.[84]

The most important aspect of the trip was the visit to Bukittingi. Fenner, Sunderland and Robson were impressed with the city, the area and its people. The first year medical students from Central Sumatra 'were an alert and enthusiastic group of young men and women.'[85] After examining the site they recommended four options for full support and limited support. The Report, however, only examined the possibility of full support. It argued that the 'key appointment' would be that of an advisory Dean, best filled by 'one individual prepared to devote much of his life to the project ... a man of energy, resourcefulness, strength of character and high ideals' and further, that two year appointments might work as an alternative suggestion.[86] In discussions with the University of California (the group assisting in the operation of the medical school in Jakarta), the authors of the Report learnt that from a pool of seventy-nine schools in the United States it had proved very difficult to recruit adequate staff. Australia, with just four medical schools, was proposing to run a much more extensive project.[87] Fenner, Sunderland and Robson pointed out that:

> Any plan of assistance on the scale proposed immediately exposes Australian medicine and Australian teaching methods, rather than individual experts, to the critical view of those already strongly wedded to other systems of education. Under those conditions it is imperative that Australia should send only its best teachers, for our efforts will need to be dramatically successful.[88]

The key problem identified in the Report was the ability of 'understaffed' Australian medical schools to man such an enterprise. Four solutions to the problem presented themselves to the authors. First, there was a need to provide support at home in order to release senior staff from Australia for visits to Bukkitingi; second, salaries must be considerably higher than those already offered in Indonesian medical schools; third, to make staffing easier block teaching would be introduced whereby anatomy, for example, could be taught in six month intensive blocks, rather than spread over a number of years; and finally, the training of suitable Indonesian staff should be prioritised, and the duration of the project should then be six to ten years.[89] Block teaching went against Sunderland's way of thinking, as he had experienced and disliked this teaching method when he was a visiting professor at Johns Hopkins University in 1954.[90]

In principle, the Report recommended that the Bukittingi project go ahead and that 'Australia should strongly back this project', providing considerable difficulties could be overcome. Fenner, Sunderland and Robson believed that the 'most valuable contribution Australia can make to Indonesia's welfare is to supply doctors trained to serve Indonesia's peculiar medical needs. This can be done only in Indonesia ... In this way Australia could make a real and lasting contribution to the welfare of the Indonesian people.'[91]

The Report had also recommended that only one medical school should run the project and by 1957, and after discussions between the Deans of all medical schools involved, it was clear that this would be Melbourne with promises of assistance from the others. In principle, agreements from the Australian and Indonesian Governments had been reached, though the Australian Government had reduced its promised finance, in the first instance, to pre-clinical training for three and a half years. Also a Dean for the new school had been found in Neil Crosby, a physiologist from Western Australia.[92] Sunderland gained preliminary agreement from informal discussions with the staff and formal agreement from the Faculty Executive at a meeting held on 2 November 1956.[93] He also returned to Indonesia with Ray and Crosby to gather further information. At the October meeting of the University Council a letter from the Department of External Affairs was tabled, asking Council to agree to the running of a pre-clinical School of Medicine at Bukittingi as a goodwill gesture from Australia to Indonesia. The Council acceded to this request, subject to negotiations on matters of detail. On 12 December the first fly landed in the ointment in the form of the Professor of Physiology, Roy Douglas 'Pansy' Wright. He wrote to Council that he believed that the matter had been 'handled in a most ingenuous and inconsiderate manner'.[94] He was furious with the secrecy of the Department of External Affairs and its announcement of the allocation of £200,000 for buildings at Bukittingi before Council approval. He lambasted Sunderland's high-handedness in not discussing the academic plans properly with the Professorial Board, and criticised all stakeholders for what he perceived to be a failure in taking into account the relative lack of success of the University of California in Indonesia. In a sustained bout of parochialism more akin to prewar Australian society, he attacked the Council for not recognising that the 'University exists by reason of an Act of the State of Victoria' meaning that all administrative arrangements, including staff contracts, would be in limbo. Most scathingly he wrote:

> The Council has committed the loyalty of the staff to an institution recognised, even at this stage, as substandard. When the actual provisions of the buildings, curricula, discipline and examinations are decided, presumably without reference to the Professorial Board or Faculty, our loyalty to the academic reputation of the University of Melbourne may require us to have nothing to do with the project.[95]

Moreover, he pointed out that the Deanship was considered to be so onerous that the holder of this office was excused from teaching. How could Sunderland possibly run Bukittingi as well? Wright had been Dean before Sunderland, and although the position was an elected one and supposed to be rotated every few years, Sunderland retired in 1971 having held the position for eighteen years.[96] Finally, to overspice the cake, Wright managed to combine his anti-Federalism with anti-Indonesian sentiments, complaining of the Commonwealth Public Service that:

> the tedious opinionated progress of a bureaucracy unenlightened as to how a university works is bad enough, but will this school be dependent on the Indonesian public service for executive action?[97]

Sunderland wrote to Council on 16 December concerning Wright's complaint, 'since his statement is in error on several points and misleading in others it calls for a reply.'[98] Sunderland pointed out that due to the delicate diplomatic nature of the project, secrecy and speed had been essential. The initial Report had also covered his concerns about the problems envisaged in setting up the school. And as for the Federal nature of the project, all could be fixed.[99] Wright's complaints fell on deaf ears and both the University and Faculty endorsed the project, and an agreement over conditions and arrangements was drawn up and settled.[100] Ray and the Anatomy Department staff were Sunderland's strongest supporters. They contributed significantly to the planning process, and all but one agreed to volunteer for teaching blocks.[101] The project was to get into full swing in 1958 but towards the close of 1957 the project was in trouble.

The postcolonial government set up by President Soekarno was increasingly moving to the left, courting China and the USSR, and becoming increasingly unpopular with the West. Furthermore McCarthyism was at its peak in the United States. In late 1957 and early 1958, CIA funded rebel elements of the newly formed Indonesian army took control of Sumatra. The rebellion was short-lived because the central government sent in loyal troops and the rebels were crushed by August 1958.

Prior to the rebel insurgency on 12 December 1957 the Melbourne *Sun* correspondent reported from 'troubled Jakarta':

negotiations with the Prime Minister, Robert Menzies.[108] Long before this took place he had become a convert to Australian involvement in South East Asia. Sunderland took every opportunity to spread his views among political powerbrokers, including the Deputy Leader of the Opposition, Arthur Calwell, at a dinner in University House in 1956. Calwell then wrote to Sunderland:

> Your views on what Australia should do for Indonesia and Asian countries in the way of establishing, or assisting in the establishment of Universities so near to our shores impressed me greatly. I very much enjoyed listening to what you had to say.[109]

For his part, Sunderland replied 'I, personally, am most grateful to you for your sympathetic interest in the good neighbour relationships which we believe should be established at a medical level with SE Asia.'[110]

A Palace on Sydney Road and doctors trained in hoards

On the 5 July 1968 Sunderland was the guest of honour at the 74th annual dinner of the Wallaby Club, a University of Melbourne walking and socialising institution. He was introduced thus:

> Much has he travelled in the realms of gold,
> And many goodly schools of medicine seen,
> In fact, his colleagues sometimes scold
> Him as the ever-absent Dean.
> Absent or not, he gets things done,
> In academic groves – and his rewards?
> A Palace on the Sydney Road,
> And doctors trained in hoards.[111]

Early in his Deanship Sunderland was faced with a new problem. Australia was growing up and Australian society, having recovered from war, was experiencing tremendous economic growth accompanied by a boom in births, coupled with a greatly increased immigration programme. According to almost all interested parties, Victoria and Australia needed more doctors. Furthermore, developments in surgery and the widespread introduction of new wonder drugs, such as antibiotics, led to a new enthusiasm for medical care. As James Le Fanu wrote:

> So dramatically successful has been the assault on disease that it is now almost impossible to imagine what life had been like back in 1945, where death in childhood ... [was] commonplace, when there where no drugs for tuberculosis, or schizophrenia, or rheumatoid arthritis, or for virtually any disease the doctor encountered; a time before open heart surgery, transplantation and test tube babies.[112]

In response to this public call for more doctors, Sunderland called for a new medical school in Victoria. He favoured a second school at Melbourne, rather than establishing a separate one at a new university. His preference for a second school, rather than expanding the present school, was because he believed that Medical Schools had a natural optimum size.[113]

In 1956 the Victorian Branch of the British Medical Association set up a subcommittee to investigate the perceived shortage of general practitioners due to many requests, almost all doubting earlier predictions that Victoria would have a surplus of doctors by 1953. The subcommittee used evidence from a variety of sources to ascertain the true number of doctors working in the community. They approached the agencies that appointed locum tenens and discovered a shortfall of thirty-five doctors. Questionnaires to all private practices further uncovered a shortfall of ninety-four, and surveys of government departments, hospitals and the like added another eighty-five.[114] This meant that in total the shortfall amounted to two hundred and fourteen doctors. Furthermore, they claimed, that the 'evidence of shortage is supported by general observation of crowded surgeries, country towns without doctors, country hospitals without resident doctors and deficiencies in services.'[115] Sir Peter MacCallum had argued against increasing the intake into the Medical School, but the subcommittee countered that 'Sir Peter MacCallum's report, like Sir Hugh Pole's survey [New South Wales], has been based on figures obtained from the medical register.'[116] The subcommittee then pointed out that this was unreliable because only initial registration was required. In conclusion, the subcommittee claimed that there was a need for one hundred and eighty new graduates each year, clearly beyond the capabilities of the one medical school in the state as constituted. In 1955 there were only one hundred and sixty-two final year graduates from the Melbourne Medical School.

In his report to the University Council on 5 May 1955, MacCallum agreed with Sunderland on the need for another medical school, but rejected the idea that the need was due to the decreasing ratio of doctors to patients in the community:

> The stark necessity of relieving an intolerable burden on those who uphold the University of Melbourne and the need to put to an end the shameful practice of admitting to its courses more students than can be given the quality of training it owes them and that they have a right to expect ... it is unjust not only to the students but to the staff and is ruinous to the University as a University.[117]

MacCallum pointed out that Sunderland had been complaining about the problems of overcrowding for years and had claimed that he 'has stated that a second Medical School will be a necessity within the next ten years. This is an understatement. It has been a necessity for at least that period already.'[118] What was the response to such pleas? According to MacCallum, 'year after year the Council indulges in an orgy of indecision ... on population figures alone we should have already two fully fledged universities and should be contemplating a third.'[119] MacCallum argued that to maintain the quality of education a good medical school should take no more than seventy students per annum, thus creating the need for another medical school.

This was problematic for Sunderland because he had earlier claimed that medical schools should remain small. In 1954 Sunderland presented the results of his trip to medical schools in Europe and North America to the University Council. He concluded that limiting the number of students was essential to maintain the quality of the course offered, thus mirroring MacCallum's position. He wrote that governments were responsible for producing adequate numbers of doctors—not universities. As for the optimum size of a medical school, Sunderland wrote 'there was ... unanimous agreement in countries where the highest standards prevail that the first year class in the medical school should not exceed 125 and should preferably be below 100.'[120] Sunderland acknowledged that such an intake

Sir Sydney Sunderland, Professor of Anatomy, 1938–61.
From the collection of the Department of Anatomy and Cell Biology.

was not feasible, but he wanted a quota of one hundred and sixty students for his school, conditional on improvements to buildings, staffing and facilities. He acknowledged the need for a selection process and advised the Council in 1954 that it should initially allow two hundred students to enrol in first year and one hundred and sixty in second year, until the selection process had settled down. Forty were expected to fail the first year going on past figures. In the early years, those who failed first year would be admitted into second year until numbers were capped in 1956.[121]

The general agreement of many stakeholders, however, was that there was a need for a second Medical School, but that it should be based at the proposed second university. Sunderland argued, when facing the prospect of a medical school free of Melbourne's control, that the clinical years should be based in the south-eastern suburbs and centred at the Alfred Hospital.[122] The Alfred was a Melbourne clinical school and relinquishing it (as Melbourne did subsequently) was, according to Sunderland, a generous gesture (not made by Sydney Medical School when another school was founded at the University of New South Wales), and one that caused significant overcrowding of the clinical years at the Melbourne School for many years after the Monash course commenced.[123] The first intake of thirty-five students at Monash did not occur until 1961. This cohort began its clinical years at the Alfred and Prince Henry Hospitals in 1964.

Prior to this intake the growing number of prospective applicants for the Melbourne Medical School had placed even greater burdens on the Anatomy Department's ability to conduct adequate dissections. As early as 1954 in a report to Council, Sunderland complained that his Anatomy Department was the worst equipped he had encountered in an extensive survey of overseas schools:

> The embalming quarters ... are inadequate for the needs of the school, are out of date, unhygienic and present a hazard to the staff compelled to work in them. The quarters are the worst seen on my visit to more than 30 schools overseas. The inconstant and uncertain supply of cadavers demands an extension of the museum facilities of the Anatomy Department.[124]

The Council had been forced to follow Sunderland's advice and cap first year entries in 1956 at two hundred.[125] Small numbers of students had been refused entry to the second year of the course from as early as 1952.[126] Sunderland was unhappy about limiting student numbers because he believed that many who missed out would have made fine doctors. No hybrid scheme, however, proved to his liking, and so the entry level was decided purely on academic scores. He was unhappy about using school results because of what he saw as great differences in the quality of tuition between schools. Thus Melbourne Medicine became the first faculty in Australia to set a quota, much to the fury of the Australian press, public, parliament and students.[127] As the *Age* told their readers, however, on 15 October 1959 under the headline, 'Victoria is Training Half the Doctors it Needs Yearly':

> A crisis in medicine, building up in Victoria, has many of our leading medical men alarmed. Authorities, including doctors, university professors, hospital officials and members of Parliament, agree that there is a shortage and that stopgap measures will no longer suffice.[128]

The newspaper report went on to state that when examining the number of registered doctors, Victoria would seem to have a ratio of one doctor to every seven hundred and sixty citizens—one of the best in the world, though the ratio seemed to the citizens of Victoria to be poor. The paradox was that:

> as medical knowledge and social conditions advance, there is ... a need for more rather than fewer doctors ... medicine has enormously expanded its boundaries and progressed in 20 years. It is far more complicated and takes far more practitioners ... our standard of living has risen ... a greater percentage are now anxious to call in their doctor and obtain the free benefits—all the new drugs and cures.[129]

And while the emergence of the biomedical approach to disease in the late nineteenth century was a pivotal event in the history of medicine, it is perhaps the period from 1940 to 1971, in which Sunderland was Professor and Dean, that ranks highest in the number and significance of medical advances. This was to prove a challenge and also fortuitous. Sunderland had to balance the two seemingly irreconcilable ideas that Victoria needed more doctors and that medical schools should remain

wanted another person to conduct all the administrative work, while he retained his research and influence within the Department. The new professor was Leslie Ray, later referred to as 'a trusted and efficient lieutenant' by Sunderland.[135]

As a further sign of Council support for Sunderland they agreed to all his terms and conditions, and added that even if he were not re-elected Dean, he could retain the salary.[136] Sunderland had bought time and Council support. Now all that was needed was funding.

In April 1962 the third and final piece was in place. The Prime Minister, Robert Menzies, offered Sunderland and businessman Kenneth Myer, part-time membership of the Australian Universities Commission (AUC). The full story concerning the changes in funding for universities cannot be included here. Briefly, the National Health and Medical Research Council (NHMRC) had been founded in 1935. By 1939 the NHMRC had dispensed about £20,000 with £18,000 going to Sydney.[137] Various piecemeal efforts were made until 1956 when the Murray Committee was convened in December to review Australian universities. It reported in December 1957 that the situation was even worse than expected, as universities were both inadequately staffed and unable to discharge their duties in relation to students or research.

'Pansy' Wright claimed that at a chance meeting in Canberra, sometime just before Menzies was contemplating the Murray option, a 'very highly placed civil servant' accosted Sunderland about why his son could not get into Medicine. Sunderland retorted that it was 'because you so and so's don't give us enough money.'[138] A combination of a revitalised economy, and state governments that were willing to let the Commonwealth dabble in what had hitherto been the preserve of the states, allowed Menzies to act. After the mid 1950s increasing numbers of students became the norm as opposed to the relatively stable levels of yearly entrants in the decade following 1945. In contrast to other faculties, however, Medicine had seen continual increases in student numbers post the Second World War. The Murray inquiry recommended increases to funding for at least another decade. As a result and by the late 1960s, Australian universities were transformed, and Sunderland was able to cash in on this boom.

In 1962 Menzies wanted a medical academic for the purposes of improving and expanding medical education. Menzies had chosen Sunderland because '[he was] suggested to me as one who would fill this latter position with great ability and I entirely agree.'[139] The AUC had the capacity to allocate large sums of money to meet specific needs in universities. Sunderland's view that an expansion of the Melbourne Medical School would be more cost-effective than starting a new medical school in the proposed Latrobe University eventually became AUC policy.[140] It is difficult not to overemphasise the importance for the Melbourne Medical School of Sunderland's position in the AUC. Improving universities was a high priority for Menzies. In fact Menzies, in replying to a letter of congratulations on his retirement from politics from Sir Leslie Martin, the chairman of the AUC along with Sunderland and other senior members wrote:

> I would like you ... to know that of the thousands of telegrams, letters and other messages I have received yours has pleased me the most; it is, in fact, the most encouraging letter I have received.[141]

The initial problem in proceeding with the expansion of the Medical School was that final AUC recommendations for the allocation of funding for the 1967–9 triennium had to be presented to the Commonwealth Government in late 1963. This meant that Sunderland had little more than a year in which to obtain agreement for his expansion plans and produce a proposal approved by all stakeholders. As AUC funding was allocated on a dollar for dollar basis along with relevant state government funding, Sunderland had to convince the Victorian Government to come up with the funds. He wrote to the Premier, Henry Bolte, on 8 August 1963 after obtaining AUC support for his plan, explaining 'I have given long and serious thought to the quickest and most economical way of providing additional places in this State for the training of doctors.'[142] He apologised for asking for an immediate agreement to match AUC funding for the project to expand the Melbourne School, but he explained that the application was due in Canberra the following week. He pointed out that:

> expanding in the manner suggested represents a radical departure from medical school policy but we are anxious to do our best to help ... This, however, is no easy exercise and the staff will be called upon to make considerable sacrifices in many ways. Despite this, I am happy to say that the plan has the unanimous backing of the Executive of the Faculty of Medicine and the support of the Acting Vice-Chancellor ... I have informed Sir Leslie

> Martin, Chairman of the Australian Universities Commission, of this plan though the decision to accept or reject it remains solely with you and your Government and the Commission will only consider the proposal at your request.[143]

It would have been hard for Bolte to refuse. Sunderland was promising half the money for a significant expansion from one hundred and sixty to two hundred and forty medical students each year. Monash would only produce thirty-five graduates in 1967 and the proposed university, Latrobe, had no voice or influence. The previous Cain Labor Government had also been sympathetic to increasing the numbers of medical graduates[144] and the Victorian Government accepted Sunderland's plan.

Sunderland was moving so fast that he had proceeded without Council approval—normally a high risk act. He was relying on the support he believed he had and this assumption proved to be correct. On 2 October Council agreed to Sunderland's proposal to increase the size of the Medical School and also agreed that 'In view of the urgency of the situation ... the Council approved of the Dean of the Faculty of Medicine informing the Australian Universities Commission of the Faculty's views before consulting the Council in this matter.'[145] Pragmatism had soundly defeated idealism regarding Sunderland's proposed ideal number of students in a Medical School.[146] And importantly, Melbourne Medical School's influence would not be compromised further by the creation of a third medical school.

As expected, Sunderland's application for AUC funding was successful and in 1964 planning for expansion began in earnest.[147] Much to the fury of many in the University, Medicine had received by far the bulk of the triennium capital grant for 1967–9. Of a total of £4,686,000, Medicine had received £3,155,000, with Arts and Education receiving just £35,000 each, or one hundredth of Medicine's allocation.[148] In Sunderland's handwritten notes taken at an unnamed meeting concerning the development of the new Medical School in the University, one sentence stands out: 'Hostility to Medicine is obvious'.[149] *Farrago* voiced many of the criticisms levelled at Sunderland's project:

> The squashing of the programme into this triennium when it was originally to be completed in stages has caused a considerable amount of speculation as

> to why and how this "acceleration" was achieved ... One of the main questions is how much does this affect the capital development of other faculties. Was it almost in the nature of a special grant, or was it made at the expense of others?[150]

In an interview with *Farrago*, Sunderland did not hide his light under a bushel. He pointed out that the Medical Faculty was: 'the apple of the governmental eye' because 'the Medical Faculty capitalised on the public demand for more doctors in this State.' He also claimed that there 'was much more general sympathy for a call for a third medical school than there is for a new Arts building.'[151] This did not allay the concerns expressed as *Farrago* continued, 'if the total grant to Melbourne University had been much smaller' due to Medicine not getting this special amount, then 'this surely reflects the paucity of the AUC grant to the University.'[152] There was another sinister aspect to the procedure. According to *Farrago* the grant application was altered in Medicine's favour after other faculties had agreed to the final submission. And the result was that 'Instead of the capital grants being spread over a number of faculties, it was concentrated on the new Medical School.'[153]

Further trouble occurred in 1966 when a financial disaster threatened the full funding of the expansion programme. Although the Medical School was now just beginning to be constructed on the southwest corner of the Parkville site, significant cuts in AUC recurrent funding meant that there was a need for stringent economic cuts throughout the University—measures that Medicine strongly resisted. The depth of the financial disaster that engulfed University administration threatened all meaningful reform encouraged by AUC funding. Medicine refused to accept the savage cuts imposed on the other faculties.[154] As a consequence of Sunderland's influence within Council, once again Medicine received preferential treatment, although this time it was required to cut its intake of students from two hundred and forty to two hundred and twenty.[155] Was Sunderland devious in his dealings over the new building? It depends upon which side of the fence one sits. In his defence, Sunderland was not the Vice-Chancellor, and his loyalties were clearly with Medicine.

On 13 June 1968 the staff 'and wives' of the Medical School were taken on a tour of the completed building, followed by a buffet supper.[156] The staff had already carried the considerable burden of planning and fitting

CHAPTER NINE

A medium sized university attached to a large medical school

IN THE 1970S IT was a common, if not ironic joke, that Melbourne was a medium sized university attached to a large medical school.[1] The move into the new medical building had meant many more students, but Sunderland kept his promise to Premier Bolte and virtually retained the same level of staffing. He claimed that if one lecturer could lecture to one hundred students they could just as easily lecture to two hundred. The same would not apply to assessment however. The expanded Anatomy Department also carried a growing postgraduate teaching load. All this had been Sunderland's idea, but in reality it was the staff of the Department that carried the extra burden.

It was serendipitous that just as the Medical School expanded, there was growing pressure on the Anatomy Department to vacate its premier position in the curriculum to make way for the expansion of medical knowledge. Both Sunderland and Wood Jones had anticipated this threat. They urged anatomists to adapt to this change by heeding the great anatomist, John Hunter, who emphasised that the study of anatomy should encompass both form and function. Anatomists, according to Wood Jones, needed to be flexible and become researchers in other disciplines, such as physiology and pathology. Heeding his own advice, Sunderland was a great advocate of interdisciplinary approaches to scientific problems. So although Anatomy had found a new home, those men who had guided the Department from the end of the Second World War to the move to the 'Palace on Sydney Road' in 1968—many of them holding appointments from the immediate postwar period—found their discipline under attack. Other problems also beset the Department. The revolution that had occurred in the public attitude to the place of women in the workforce had to be accommodated. No woman had ever been appointed to a permanent academic position in the Anatomy Department (apart from Bernice Stratford in 1965) when the new building was occupied in 1968, although female medical graduates had been around for over eighty years. And few, if any, postgraduates in the Department were female. With the increase in funding for universities in the 1960s, a new research-based culture had emerged that equated success with the amount of research money won. The research tradition of the Department had been to work alone or in small groups (after the teaching had been done) and not in highly organised and generously funded laboratories. New and expensive technologies, such as the electron microscope, had become essential research tools and were scarcely affordable under the old model. The new building and the sudden increase in numbers had also taken up much of the Department's energy, thus staff had failed to attract the growing amount of research funding available in the 1960s. In this new milieu, gross anatomy and physical anthropology were endangered species, and this was to be a major theme in the Department's research history. As John Furness, Professor and Head of the Department (1993–96), and again in 1999, wrote:

> Progressively from the 1960s, and with accelerating pace in the 1970s and 1980s, the place of Topographic [Gross] Anatomy in the medical curriculum came under siege. Similar pressures were encountered in the teaching of Topographic Anatomy within dentistry, physiotherapy and other health-related professional degrees. The major reason was that although the length of these degrees had remained stable, more-or-less, for decades, the knowledge base on which modern medicine and health practice was based had expanded enormously. Medical Schools were faced with the problem of fitting into their curricula emerging areas of immunology, molecular biology, pharmacology and therapeutics, epidemiology, virology, new disease entities and new treatments. The numbers of hours devoted to teaching anatomy, and the need to learn the fine details of anatomy began to be questioned,

> and it was even suggested that traditional teaching around the body in a dissecting room should be abandoned in favour of teaching from specimens and models, or entirely from computer-based instruction programs.[2]

In 1971 Sunderland retired as Dean to his research laboratory, to which he remained a daily visitor almost until his death in 1993. Les Ray continued as Head of Department until 1978.

When Keith Bradley's chair became vacant in 1977 Graeme Ryan, a pathologist who was leading a laboratory in the Pathology Department and working on renal disease, was approached by the Dean, Lance Townsend, and his Deputy (and later successor) David Penington, and invited to take Bradley's Chair, as well as head the Department. Ryan, a Melbourne medical graduate, had worked at St Bartholomew's Hospital in London, with Guido Majno, on inflammation and repair in Switzerland, and then at Harvard University, with Morris Karnovsky, on renal disease using electron microscopy. He returned to Melbourne after nearly eight years absence and due to a strange sequence of events, became Professor of Anatomy in 1978. He also brought a large grant from the NHMRC, as well as his laboratory, into the Department. At this time the Department was at the crossroads, with changes in the Medical School's direction already afoot. Many new disciplines wanted a slice of the curriculum and departments were increasingly under pressure to conduct more research, and this well suited Ryan's appointment. His emphasis on research, as well as interests unconnected to gross anatomy, meant he would take the Department in a new direction. This was a critical moment in the Anatomy Department's history. All succeeding departmental Heads date the beginning of the current Department from this time. The current Head, Antony Goodwin, also noted that the demarcation between anatomy and physiology was quite distinct, and contrary to the spirit of Sunderland and Wood Jones's utterances. The changes subsequently initiated by Ryan expanded the credentials of the Department as a research centre. Ian Darian-Smith, who was already holding a Chair in Physiology, was appointed Professor and Deputy Head of the Department in 1982 after Ray's unexpected death. He became Head of Department when Ryan was appointed Dean in 1986. Darian-Smith's legacy was to combine all neurological courses in Physiology and Anatomy into a single neuroscience stream; a very significant part of the Medical School today.[3]

He remembers being welcomed into the Department and the departmental support for his different approach, ensuring his most productive work period. He strongly believes that the Department has a rich and certain future, especially as the great advances in the neurosciences will allow researchers to return to the mind and body as it operates on a macro level. Darian-Smith retired as Head in 1992, but continued working in his laboratory. Thus the transformation was complete.

Today the Department's research activities focus on three major themes in a number of laboratories: neuroscience, cell and developmental biology and functional and clinical anatomy. The scope of activities range from molecular biology to cell culture, electrophysiology, invertebrate and vertebrate development, behavioural research, anatomical studies of the human neurovascular and lymphatic systems, and educational research centred on the development and evaluation of innovative multimedia learning aids for students of Human Anatomy (*Anatomedia*) and Histology and Pathology (*Patient Under the Microscope*). The expansion of research means that in 2007 there are more researchers in the Department than teaching academics.

Increased research funding has meant that there have been significant improvements in the physical resources of the Department, including the new museum and modern laboratories. There has also been public outreach in the form of many permanent displays in the Mind and Body Gallery produced for the Melbourne Museum.

Although it had been a long wait for women, the Department soon began reforming itself after moving to the new building. The first female permanent academic member of the Anatomy Department was Bernice Stratford, who graduated in 1950 and specialised in obstetrics. After raising her children she worked in the Pathology Department under Edgar King, gaining a PhD in Histology, and taking up a senior lectureship in Histology and Embryology in the Anatomy Department from 1965. Stratford recalls that it was some years before her older male colleagues deemed that she was capable of teaching medical students, instead of dental and physiotherapy students. She remained in the Department into the 1980s and took early retirement to teach in China.[4] Her appointment seemed to break the ice and coupled with the growing need for more staff in the new building, a number of medically qualified women married to medicos were soon demonstrating in the Department. They were christened the 'medical mums', including Mary Wheeler, Joan Towns, Marie Cockbill, Lynn Billings and physiotherapist, Coralie Kenny. Although they held part-time positions these women brought a new atmosphere and enthusiasm to the Department. And not long after their appearance there was a growth in the number of full-time postgraduate women who took up places. The first full-time, non-medical trained member of staff was Ros Perry in Histology. Another member of this group of postgraduate women, Daine Alcorn, went on to become the first female Professor of Anatomy in the Department (1996–2002) and Head of the Department (1996–8, 2000–2). Her research was on the development of the foetal lung and she also worked with Ryan in his renal laboratory. In July 2002 Alcorn left the Department to become Dean of the Faculty of Life Sciences at RMIT University (and later Pro Vice-Chancellor) and Antony Goodwin was appointed Head, a position he currently holds. Goodwin's original degree was in electrical engineering from Witswatersrand University in South Africa (Keith Bradley had also started his career as an engineer). He moved into medical research, completing a PhD in biological control systems at Caltech in California before moving to the Australian National University in Canberra under Peter Bishop. He moved to Melbourne in 1974 to work in Darian-Smith's laboratory and followed him to the Anatomy Department. Goodwin was awarded a Personal Chair in 2001. He was Deputy Head of Department before he was appointed Head in 2002. His interests are in neural processing, particularly sensorimotor function of the hand, thus adding to the neurosciences strength in the Department.

Collections

The most difficult recent legacy from the pre-1980s Department has been the Berry and Murray Black collections of skeletal remains, mostly of Aboriginal origin. This book does not deal with this topic in detail because it is far too complex to summarise. Briefly, the remains were collected from the very beginning of the Department's history until approximately the end of the Sunderland and Ray era. During that time relatively little was published on the remains and certainly nothing of significance. Partly as a consequence of a change in public opinion in the 1970s, when many came to believe that it was inappropriate for an Anatomy Department to hold such collections, they were sent to the Museum of Victoria for repatriation. As a result, the Department no longer holds any such material.

The first collectors of skeletal remains were Halford and Berry, using the skulls to provide evolutionary models that created a hierarchy of races. Wood Jones and Sunderland also possessed a keen interest in physical anthropology. Many of the earliest investigators were biological racists. Wood Jones did not hold such beliefs, although he too was a keen collector and investigator of Aboriginal crania, and was a strong public supporter of Aboriginal equality.

Colin McKenzie also deserves a mention as a keen anthropologist who has the distinction of convincing the Commonwealth Government to pass the *Institute of Anatomy Act 1924* to establish an Institute of Anatomy in Canberra for the purpose of studying physical anthropology—an indication of the perception of the discipline's importance at the time. McKenzie moved his extensive collection to the Institute, thus making the Anatomy Department at Melbourne the parent of the forerunner of the Australian National University.

By Sunderland's era some material collected by Halford, Berry, McKenzie and others had been dispersed, and the remaining material was divided into the Murray Black and Berry collections. Ray had stewardship of the Murray Black collection and Russell, the Berry collection. After Ray's death and Russell's retirement the collections had no place in the research profile of the Department. Thus their repatriation was as much a consequence of the changing research focus of the evolving Department as it was of changes in community attitudes.

Gordon Keys Smith worked on a project with the collection under the direction of Sunderland just after the war. He wrote in his memoirs that the exercise was poorly constructed with little recorded data concerning the specimens:

> As far as I know, the project was never completed and my measurements, with those of other staff, are probably in some cupboard somewhere, if they have not been lost altogether. Now it is realised that such a collection of skeletons is offensive to Aborigines and I would hope that long ago these bones have been returned to their Aboriginal communities.[5]

Anatomy today

On 6 May 2006 the *Australian* published a front page headline, 'DOCTORS FAIL BASIC ANATOMY', followed by an article that raised the concerns of senior clinicians about the lack of knowledge of anatomy among medical school graduates. Senior consultants were 'horrified' that some graduates did not know where the prostate gland was located, according to the newspaper's sources. Chris Briggs, Deputy Head of the Department, takes up the story:

> Until the early 1970s Anatomy was still one of the pillars of the medical program (together with Physiology, Biochemistry and Pathology) although change was in the air. At the time, a total of 600 hours was devoted to Anatomy (including Topographic Anatomy, Histology, Embryology and Neuroanatomy), taught in the second and third years of the course. I suspect this total would not have varied greatly from the early years. The first major change to the medical curriculum was introduced in 1974, and later modified in 1978. This involved eliminating much of the content of first year Medical Biology, which the students were becoming bored with because of the perceived repetition of Year 12 material, thereby freeing up space to move preclinical teaching to the first and second years. There was also a need to make room for growth subjects such as Immunology and Pharmacology. Anatomy teaching came out of these changes intact. It was still a very big subject comprising about 312 contact hours, exclusive of Histology, Embryology and Neuroanatomy. Students dissected the entire body in groups of about eight per table, and also had weekly tutorials conducted by full-time and part-time demonstrators (all surgical trainees) with written and viva voce examinations at the end of second year.
>
> This program continued until 1999 when the Faculty implemented a move to a problem-based independent learning curriculum, modelled in part on similar programs that were developing elsewhere in the world. Anatomy, together with other preclinical subjects, disappeared as discrete entities and became part of semester-long systems-based units. Initially, Anatomy teaching (excluding Embryology, Histology and Neurosciences) was significantly reduced, although this has levelled off at about 124 hours each year. Anatomy lectures are interspersed throughout each of the semester-long

units and the students have practical classes, incorporating Radiology, Osteology and wet specimen workshops. Dissection has been reduced and now comprises dissection of selected regions (limbs, thorax, abdomen and pelvis), as well as the anatomical basis of common practical procedures required of a first-port-of-call medical practitioner, such as intercostal catheterisation and venepuncture among others.

In 2000 the Department perceived an opportunity to fill a void in postgraduate Anatomy instruction for surgical trainees, many of whom were graduating from medical schools in which anatomy teaching had been pared to the bone. The Department established the Diploma of Surgical Anatomy in conjunction with the then Melbourne University Private, and now the University School of Enterprise, and the Royal Australasian College of Surgeons. To date, more than 100 candidates have availed themselves of this advanced anatomy training.

The current members of the Department (many of them are not anatomists) believe it is well worth investing in anatomy teaching for the future, and are pleased that they have retained a strong cohort of anatomists, and a good bank of anatomical know-how. International trends also indicate that Anatomy is making a comeback. Anatomy teaching in the Department currently thrives under Chris Briggs and Norm Eizenberg through their strong link with the Royal Australasian College of Surgeons, and through the ongoing creation of *Anatomedia*, a computer-based anatomy course used in many universities and recently highly recommended by the *Lancet*—a far cry from its review of Halford's anti-Darwinian tract in the 1860s. The forensic work of Briggs, which takes him to catastrophes such as the Bali bombings, further enhances the profile of anatomy teachers. With such an optimistic prognosis, the problem the Department faces is the recruiting of anatomy teachers for the future. With the dearth of anatomists graduating from other universities, and the inability of Topographical Anatomy to provide a research career, positions are hard to fill. Young surgeons find it difficult to provide voluntary teaching due to the high costs of setting up practice, including medical negligence insurance. Goodwin and Briggs both see a bright future for the Department with the growth of Radiology and all the new investigative procedures, but they argue that the University and the Medical School need to be aware of recruitment problems and assessment of non-research academic staff. There is also a current debate about methods of anatomical teaching, with a move to incorporate new technologies to replace the time currently spent on dissecting cadavers.

The change in name to the Department of Anatomy and Cell Biology in 1993 was part of the reforms begun in Ryan's time, and better recognises the shift from the pure topographical anatomical character of the old Department to the new focus on cell structure and function. There was opposition from without the Department, but with the cooperation of Botany and Zoology this area now constitutes one of the Department's significant teaching streams.

The group who attended the first dissection in the Medical School in 1863 in Halford's backshed in his Swanston Street home were an eclectic mix, and one that did not include even one self-proclaimed anatomist.[6] They consisted of a physiologist; a gynaecologist; a poet of very considerable talent and a future physician; a polymath who was a theatre critic, forensic pathologist, acting coroner and lecturer in forensics; a future demonstrator in the Department; and a future Presbyterian minister. The Anatomy Department has never been a slave merely to dissection and mapping the human body. It has housed many men and women of high talent with an extraordinary range of interests. As Wood Jones told his audience at Adelaide University in 1923, an anatomist was not just a cabman that knows the streets of the city in which he plies for hire, but an anatomist 'is the father of the physiologist, of the embryologist, the histologist and the anthropologist, among others.'[7] Within the Department today this list would need to be greatly extended. It is precisely this diversity of interests and a willingness to adapt that has created a healthy and vital Department in the University of Melbourne, and also led to a fascinating history. Would the previous members of the Department recognise it today? They most certainly would.

Notes

INTRODUCTION

1. See generally Bryan Gandevia, *The Melbourne Medical Students: 1862–1942*, The Melbourne Medical Students' Society, Melbourne, 1948, p. 40.
2. *Speculum*, no. 1, July 1884, p. 1. The Commencement also doubled as the degree conferring ceremony. There was good cause to claim they created the greatest disturbance.

CHAPTER ONE (PAGES 1–28)

1. Robert Lacey describes the Duke (then King George V) as such: 'as his official biographer felt compelled to admit, King George V was distinguished 'by no exercise of social gifts, by no personal magnetism, by no intellectual powers. He was neither a wit nor a brilliant raconteur, neither well-read nor well-educated, and he made no great contribution to enlightened social converse. He lacked intellectual curiosity and only late in life acquired some measure of artistic taste.' He was, in other words, exactly like most of his subjects. He discovered a new job for modern kings and queens to do: representation.'
2. *Alma Mater*, vol. ii, no. 1, April 1897, p. 17.
3. Ibid, vol. 6, no. 6, June 1901, p. 16.
4. For an example of the opposition see William Thomson versus Brownless and the Medical School in the *Argus*, 4 April 1864, p. 6; 7 April 1864, p. 6; 8 April 1864, p. 6. For Thomson's ongoing dispute with the University see K. F. Russell, *The Melbourne Medical School, 1862–1962*, Melbourne University Press, Carlton, 1977, pp. 42–5. For the supply of doctors see Harold Love, *James Edward Neild, Victorian Virtuoso*, Melbourne University Press, Carlton, 1989, p. 85; T. S. Pensabane, *The Rise of the Medical Practitioner in Victoria*, Health Research Project, Research Monograph 2, Australian National University, Canberra, 1980, chapter 4.
5. K. F. Russell, The Melbourne Medical School, p. 11.
6. Thomas A. Darragh, 'Frederick McCoy: the Irish Years', *The Victorian Naturalist*, vol. 118 (5) 2001, McCoy Special Issue, part 1, p. 160.
7. ibid., p. 163.
8. University of Melbourne Calendar, 1857, 'General Examinations December 1856, Comparative Anatomy', p. 63.
   '1. Contrast the circulation of the blood in reptiles.
   2. What is the normal construction of the heart in the Mammalia, Aves, Reptilia, and Pisces respectively?

3. How are the nervous centres of the Articulata arranged?
4. Describe the characters and relative positions of the cement, dentine, and enamel in the teeth of mammalia.
5. By what structural characters might you refer a minute fragment of bone to the classes of Mammals, Birds, or Reptiles, respectively?
6. By what characters could you distinguish a human blood stain, from one produced by the blood of birds or quadrupeds?
7. Enumerate the bones of the anterior and posterior extremities in the higher Vertebrata.
8. Describe the course of the alimentary canal in the Lamellibranchiata.
9. Describe the various modes of respiration in the Mollusca.
10. Of what bones is the carapace of the Chelonia formed, and what modifications have they undergone?'

9. For Cuvier see Ann Mozley, 'Evolution and the Climate of Opinion in Australia, 1840–76' *Victorian Studies*, June 1967, vol. 10, pp. 412, 425. For the textbooks see *University of Melbourne Calendar 1858–9*, 'texts: Owen's Lectures of Comparative Anatomy & Zoology, Carpenter's (ed. 1851) Cuvier's Animal Kingdom, De la Beche's Geological Observer', p. 46.
10. Harry Brookes Allen Papers, University of Melbourne Archives, Box 63/3, lecture notes of Comparative Anatomy, 1872.
11. *Alma Mater*, vol iii, no. 6, September 1898, 'Brownless Memorial Supplement, Being a Special Medical Supplement', p. 9. For McCoy's sad end as a lecturer see Ian Wilkinson, 'Frederick McCoy and the University of Melbourne' *The Victorian Naturalist*, vol. 118 (5) 2001, McCoy Special Issue, part 1, p. 189.
12. University of Melbourne, Medical School Jubilee 1914, Ford, Carlton, 1914. Professor Sir Harry Allen, 'A History of the Medical School', p. 17.
13. W. A. Osborne, 'George Britton Halford: His Life and Work: The Halford Oration' *Medical Journal of Australia*, 19 January 1929, p. 66.
14. Harry Brookes Allen Papers, University of Melbourne Archives, Box 63/3, notebook on General Anatomy, Physiology and Pathology, vol. II, 1873.
15. Charles Kingsley, *The Water Babies*, Dent, London, 1957 (1863), p. 58.
16. E. Graeme Robertson, 'Melbourne's Public Anatomical and Anthropological Museums, and the Jordans', *Medical Journal of Australia*, 4 February 1956, p. 164.
17. ibid., p. 166.
18. *Argus*, 27 December 1860.
19. See Paul Turnbull, 'Rare work amongst the professors: the capture of indigenous skulls within phrenological knowledge in early colonial Australia' in Barbara Creed and Jeanette Hoorn (eds), *Body Trade: Captivity, Cannibalism and Colonialism in the Pacific*, Pluto Press, Australia, 2001.
20. Cesare Lombroso, *Criminal Man*, Putna, 1911.
21. *Argus*, 9 November 1912.
22. Nancy Stepan, *The Idea of Race in Science: Great Britain, 1800–1960*, Macmillan, London, 1982, p. 28.

23. Reverend Ralph Brown, *A Delineation of the Character, Talents, Physiological Developments and Natural Adaptations of M.*, Melbourne, 1894, pp. 24–5.
24. R. Brough Smyth, *The Aborigines of Victoria*, Ferres, Melbourne, 1878, pp. 338f.
25. *Illustrated Melbourne Post*, 20 December 1862.
26. Contrary to rumour, the Anatomy Department museum does not hold the original skull.
27. Reminiscences in Mary Kehoe, *The Melbourne Benevolent Asylum: Hotham's Premier Building: the Annals of Hotham, volume 1*, Hotham History Project, North Melbourne, 1998, pp. 18–19.
28. F. W. Nicholas and J. M. Nicholas, *Charles Darwin in Australia*, Cambridge University Press, Cambridge, 2002, p. 140.
29. R. A. Cage, *Poverty abounding, charity aplenty: the charity network in colonial Victoria*, Hale & Iremonger, Sydney, 1992. For the itinerant nature of the population see Ross L. Jones, 'The Rogers-Templeton and Pearson Royal Commissions: Contemporary Views of the 1872 Education Act', *History of Education Review*, vol. 27, no. 2, 1998, pp. 50–66.
30. For example, in 1869, three hundred and sixty-nine men and one hundred and fifty-five women, *Argus*, 30 July 1869 p. 5.
31. Mary Kehoe, *The Melbourne Benevolent Asylum*, p. 19.
32. Warwick Anderson, *The Cultivation of Whiteness: Science, Health and Racial Destiny in Australia,* Melbourne University Press, Carlton South, 2002, p. 21.
33. The infamous lunatic Asylum in London.
34. *Argus* supplement, 10 July 1867, p. 3.
35. Richard Stawell, 'The Foundation of a Medical School and the Progress of Medical Education: the Halford Oration' *Medical Journal of* Australia, 3 January 1931, p. 3; H.B. Allen, *University of Melbourne, Medical School Jubilee* 1914, Ford, Carlton, 1914, p. 5.
36. David M. O'Sullivan, 'David John Thomas, FRCS, A Founder of Victorian Medicine', *The Medical Journal of Australia*, June 30, 1956, pp. 1078–9.
37. K. F. Russell, *The Melbourne Medical School,* p. 46.
38. David M. O'Sullivan, op. cit., p. 1083.
39. K. F. Russell, *The Melbourne Medical School,* op.cit., p. 30.
40. David M. O'Sullivan, op. cit., p. 1083.
41. University of Melbourne, *Medical School Jubilee 1914*, pp. 97–8.
42. T. S. Pensabane, *The Rise of the Medical Practitioner in Victoria*, op.cit., p. 58.
43. 'The Profession in Victoria', *Medical Examiner,* 5 April 1877, p. 274, cited in T.S. Pensabane, loc.cit.
44. James Edward Neil, 'The Medical School of the Melbourne University: part II', *Australian Medical Journal*, 15 June 1887, vol. IX, no. 6, pp. 242–3.
45. T.S. Pensabane, loc.cit., p. 69.
46. *Medical School Jubilee 1914*, pp. 107–8; T.S. Pensabane, loc.cit.
47. W. A. Osborne, 'George Britton Halford: His Life and Work: The Halford Oration' *Medical Journal of Australia*, 19 January 1929, p. 65.

48. Barry Butcher, 'Gorilla Warfare in Melbourne: Halford, Huxley' and 'Man's Place in Nature' in Home, R.W. (ed.), *Australian Science in the Making*, Cambridge University Press, Cambridge, 1988, pp. 166–7.
49. The first dispute in 1863 is discussed in the following chapter.
50. *Argus,* 20 June 1864, p. 6.
51. *Times,* 18 October 1864, p. 9.
52. *Argus*, 19 July 1864, p. 7.
53. For Macewen's primacy rather than Bennett and Godlee see Malcolm Macmillan, 'Localisation and William Macewen's Early Brain Surgery Part I: The Controversy', *Journal of the History of the Neurosciences*, vol. 13, no. 4, December, 2004; Malcolm Macmillan, 'Localization and William Macewen's Early Brain Surgery Part II: The Cases', *Journal of the History of the Neurosciences*, vol. 14, no. 1, March 2005, pp. 24–56.
54. *Argus,* 23 July 1864, p. 7.
55. Harold Love, op.cit., p. 121.
56. *Leader*, 6 August 1864, p. 1.
57. ibid.
58. For example, the *Age*, 5 August 1864, p. 5; *Herald*, 5 August 1864, p.2; *Illustrated Melbourne Post,* 20 August 1864, p. 1.
59. *Illustrated Melbourne Post*, 20 August 1864, p. 3.
60. ibid.
61. ibid.
62. *Times,* 18 October 1864, p. 9.
63. *Illustrated Melbourne Post*, loc.cit.
64. *Argus,* 6 August 1864, p. 5.
65. *Melbourne Punch*, 11 August 1864, p. 52.
66. R. J. W. Selleck, *The Shop: The University of Melbourne: 1850–1939*, Melbourne University Press, Carlton, 2003, pp. 108–9.
67. *Argus* supplement, 10 July 1867, p. 3.
68. *Age*, 29 June 1869. Huxley's protoplasmic hypothesis was anti-vitalist, that is, he believed the chemistry of the cell was the same as inorganic chemistry, there was no 'spiritual' force animating it. He wrote, 'the living body obeys the mechanical laws of ordinary matter.' For Halford see R. J. W. Selleck, *The Shop*, p. 106.
69. R. J. W. Selleck, op.cit., pp. 382–4.
70. W. A. Osborne, op.cit., p. 68.
71. Alan Newton, 'On Surgical Education: the Halford Oration', *Medical Journal of Australia*, 9 January 1937, p. 42.
72. E. Graeme Robertson, 'Melbourne's Public Anatomical and Anthropological Museums, and the Jordans', *Medical Journal of Australia*, 4 February 1956, p. 168. The BMA Medical Reform Committee successfully secured the passing of the *Medical Act 1858*. The Act established the General Medical Council and the Medical Register, therefore distinguishing, for the first time, between the qualified and unqualified practitioners.

73. T.S. Pensabane, *The Rise of the Medical Practitioner in Victoria*, op.cit., p. 7.
74. C. Craig, 'The Egregious Dr Beaney of the Beaney Scholarships', *Medical Journal of Australia*, 6 May 1950, pp. 593–8; Bryan Gandevia, 'Some Aspects of the Life and Work of James George Beaney', *Medical Journal of Australia*, 2 May 1953, pp. 614–9.
75. Bryan Gandevia, 'Some Aspects of the Life and Work of James George Beaney', p. 614.
76. ibid., p. 615.
77. J. G. Beaney, *Doctors Differ: A Lecture*, Melbourne, 1876, pp. 23–4. Quoted in T.S. Pensabane, loc.cit.
78. For the figures of the 1865 and 1875 elections see Bryan Gandevia, 'James Thomas Rudall, FRCS (1828–1907): His Life, and Journal for the Year 1858', *Medical Journal of Australia*, 25 December 1954, p. 992.
79. C. Craig, 'The Egregious Dr Beaney of the Beaney Scholarships', p. 594.
80. David M. O'Sullivan, 'David John Thomas, FRCS, A Founder of Victorian Medicine', *The Medical Journal of Australia*, June 30, 1956, p. 1081.
81. See Janet McCalman, *Sex and Suffering: Women's Health and a Women's Hospital: The Royal Women's Hospital, Melbourne 1856–1996*, Melbourne University Press, Carlton South, 1998, especially chapters 1 and 2.
82. F. B. Smith, 'Disputes about Typhoid Fever in Victoria in the 1870s', *Health and History, Journal of the Australian Society of the History of Medicine*, vol. 4, no. 2, 2002, pp. 1–18; K. F. Russell, *The Melbourne Medical School, 1862–1962*, MUP, Carlton, 1977, pp. 42–5.
83. R. J. W. Selleck, *The Shop*, op.cit., pp. 106–7.
84. K. F. Russell, *The Melbourne Medical School*, pp. 52–3; *Argus* 25 October 1869, pp. 4, 7; 27 October 1869, pp. 5, 6.
85. K. F. Russell, op. cit., p. 60.
86. ibid.
87. *Argus*, 25 September 1867, p. 5; E. Graeme Robertson, 'Melbourne's Public Anatomical and Anthropological Museums, and the Jordans', *Medical Journal of Australia*, 4 February 1956, p. 170.
88. E. Graeme Robertson, op. cit., p. 169.
89. ibid.
90. ibid., p. 170.
91. ibid.
92. ibid.

CHAPTER TWO (PAGES 29–46)

1. 'The Profession in Victoria', *Medical Examiner*, 5 April 1877, p. 274 cited in T.S. Pensabane, *The Rise of the Medical Practitioner in Victoria*, p. 58.
2. *Medical School Jubilee 1914*, p. 51.
3. For a detailed account of the debate see Barry W. Butcher, 'Gorilla Warfare in Melbourne'. Also Barry W. Butcher, 'Darwinism, Social Darwinism, and the

Australian Aborigines: A Reevaluation' in MacLeod, Roy and Rehbock, Philip (eds), *Evolutionary Theory and Natural History in the Pacific: Darwin's Laboratory*, University of Hawaii Press, Honolulu, 1994. Barry W. Butcher, Darwinism and Australia: 1836–1914, unpublished PhD thesis, University of Melbourne, 1992, p. 74. Butcher has shown the degree to which Darwinism quickly became the intellectual paradigm in the Australian colonies.

4. W. E. Hearn, *Plutology, or the Theory of Efforts to Satisfy Human Wants*, Melbourne, 1863; R. J. W. Selleck, *The Shop*, pp. 37–40.
5. See Adrian Desmond, *Huxley, The Devil's Disciple*, Michael Joseph, London, 1994, pp. 262–3.
6. Stephen Jay Gould, *Bully for Brontosaurus: Reflections in Natural History*, Vintage, London, 2001, pp. 309–24.
7. The popular myth that claimed victory for Huxley at that debate is probably wrong. See J. Vernon Jensen, 'Return to the Wilberforce-Huxley debate', *British Journal for the History of Science*, vol. 21, 1988.
8. Charles Kingsley, *The Water Babies*, Dent, London, 1957 (1863), pp. 126–7.
9. R. J. W. Selleck, op.cit., pp. 85–93.
10. Joan M. Dixon, 'Melbourne 1865: Gorillas at the Museum', in Carolyn Rasmussen, *A Museum for the People: A History of Museum Victoria and its predecessor institutions, 1854–2000*, Carlton North, Scribe Publications, 2001.
11. *Argus*, 20 June 1865.
12. Ibid, 25 July 1865, p. 6.
13. R. J. W. Selleck, op.cit., pp. 91–3.
14. *Argus*, 20 July 1863, p. 4.
15. ibid. In seeming contradiction to its normally conservative stance the *Argus* was, from the first, generally pro-Darwin. This was essentially due to the role of its founder and first editor Edward Wilson who admired Darwin and, after early retirement, was a close neighbour in Kent and provided Darwin with Australian material for his studies. The paper's editorial policy wavered after his departure but would have also been strengthened by the opposition of the *Age* newspaper to Darwinism. The *Argus* invariably disagreed with the *Age* on all matters. For an extensive and instructive discussion on this topic see Barry Butcher, 'Gorilla Warfare in Melbourne', pp. 158–60.
16. *Melbourne Punch*, 23 July 1863, p. 31.
17. ibid. Probably written by Neild.
18. W. A. Osborne, 'George Britton Halford, p. 68.
19. G. B. Halford, *Not Like Man, Bimanous and Biped; Nor Yet Quadrumanous; but Cheiropodus*, Wilson & Mackinnon, Melbourne, 1863; G. B. Halford, 'On the skeleton of the gorilla', *Transactions of the Royal Society of Victoria*, vol. 7, 1866; G. B. Halford, 'Quadrumanous or bimanous?', *Australian Medical Journal*, July 1863; G. B. Halford, 'The terminal divisions of the limbs of man and monkey', *Argus*, 17 July 1863.

20. Arthur O. Lovejoy, *The Great Chain of Being: A Study of the History of an Idea*, Harper, New York, 1960, p. 229.
21. *The Lancet*, 12 December 1863, p. 681.
22. ibid., p. 683.
23. ibid.
24. W. A. Osborne, op.cit., p. 69.
25. Adrian Desmond, op.cit., p. 341.
26. Serle, G. and Nede, B. (eds), *Australian Dictionary of Biography, 1891–1939*, Melbourne University Press, Carlton South, 1979–90; F. B. Smith, 'Disputes about Typhoid Fever in Victoria in the 1870s', *Health and History: Journal of the Australian Society of the History of Medicine*, vol. 4, no. 2, 2002, pp. 1–18; Warwick Anderson, *The Cultivation of Whiteness,* pp. 48f.
27. *Argus*, 22 July 1863, p. 5.
28. Quoted in Barry Butcher, 'Gorilla Warfare in Melbourne', p. 166.
29. James Barrett, 'The Halford Oration', *Medical Journal of Australia*, 10 December 1932, p. 708.
30. A. L. Kenny, 'Halford the Man: The Halford Oration', *Medical Journal of Australia*, 1 February 1941, p. 130. W. A. Osborne agreed, see W. A. Osborne, 'George Britton Halford', p. 69.
31. There were four major objections to Darwin's theory:
    1. St George Jackson Mivart's argument that 'there must have been a stage at which the incipient organ [for example, the eye] had no recognizable function, and would therefore have conferred no selective advantage'.
    2. If evolution is the product of small changes over a huge expanse of time, where are the intermediate types in the fossil record?
    3. Lord Kelvin's evidence that the earth wasn't old enough to allow time for evolution (he did not know about radiation and thus thought the world was cooling fast).
    4. Fleeming Kenkin exposed Darwin's ignorance of the mechanism of inheritance. He pointed out that favourable variation would soon disperse itself equally through the population as it became 'diluted' through breeding in successive generations—Darwin had no answer.
32. Harry Brookes Allen Papers, University of Melbourne Archives, Box 63/3, loose sheet in notebook on General Anatomy, Physiology and Pathology, vol. II, 1873. First, the highest types of the classes often appeared first, for example, as had the Placoida and Ganoida of fish in the Silurian and Devonian period. Also Dinosaurs were found in the Wealden clay while snakes did not appear until the Tertiary. Second, there was an absence of any links between classes, orders or families, for example, reptiles and birds. Third, means of species always appeared before extremes. Thus Pleisaura existed in Oxford clay before either Ichthyosaurus or Plesiosaurus. Fourth, there was an absence of any regular graduation in the fossil record. Fish and Reptiles have teeth, so do mammals, while birds have none. Fifth,

some organs are peculiar to certain classes. Thus the optic tubercules are found in fish only, but are absent equally in the lower and higher fish, thus showing no evidence of evolution. Sixth, the highest types of a species are always the first created, thus giants of a type appeared before the smaller versions.

33. Adrian Desmond, op.cit., pp. 262–3. While the late 1860s and 1870s were the high point of Darwinism, later in the century his theory of natural selection came under serious attack again, though not from those arguing for divine intervention. Biologists began to argue that direction was a product of the genetic constitution of species.
34. S. V. Sewell, 'The Evolution of the Language Mechanism', in *Transactions of the British Medical Association Congress*, 1923, Section VII, Neurology and Psychiatry, supplement to the *Medical Journal of Australia*, June 7, 1924, p. 391.
35. Peter J. Bowler, *Evolution: The History of an Idea*, University of California Press, Berkeley, 1989, pp. 196–7.
36. *Age*, 22 April 1882.
37. See Jill Roe, 'Challenge and Response: Religious Life in Melbourne, 1876–86' in *Journal of Religious History*, vol. 5, 1968/9, pp. 149–66.
38. In Adrian Desmond and James Moore, *Darwin*, Penguin, London, 1991, p. 677.
39. Nettie Palmer, *Henry Bournes Higgins: A Memoir*, George Harrap, London, 1931, p. 74.
40. Adrian Desmond, op.cit., pp. 367–9.
41. *Age*, 29 June 1869. Huxley's protoplasmic hypothesis was anti-vitalist, that is, he believed the chemistry of the cell was the same as inorganic chemistry; there was no "spiritual" force animating it. He wrote, 'the living body obeys the mechanical laws of ordinary matter.'
42. *Argus,* 20 June 1865.
43. R. J. W. Selleck, op.cit., p. 106.
44. *Leader,* 3 July 1869, p. 13.
45. ibid. See also *Argus,* 2 July 1869, p. 6; *Argus,* 25 July 1869, p. 5; *Argus,* 31 July 1869, p. 5.
46. *Leader,* 3 July 1869, p. 13.
47. R. J. W. Selleck, op.cit., p. 287.
48. See early discussions in the *Proceedings of the Zoological and Acclimatisation Society of Victoria, 1862–73*, vol. i, Masterman, 1872; vol. ii, Stillwell and Knight, 1873; vol. iii, Stillwell and Knight, 1874; vol. iv, Sands and McDougall, 1875; vol. v, Sands and McDougall, 1878, Melbourne.
49. *Proceedings of the Zoological and Acclimatisation Society of Victoria, First Report, 1862*, vol. 1, unnumbered pages of membership lists at the beginning; pp. 34–5; p. 19; *Second Report*, 1863, p. 7; *First Report, 1862*, p. 9; *Argus*, 30 August 1928, p. 12.
50. Warwick Anderson, *The Cultivation of Whiteness*, pp. 23f.
51. ibid., p. 11. Anderson has established the importance of this strand of thought in medical Australia in the first half of the book.

52. *Herald*, 5 September 1934. See Warwick Anderson, op.cit., pp. 27, 62f.
53. *Speculum,* no. 23, February 1891, pp. 161, 162.
54. ibid, no. 30, August 1894, p. 19.
55. *Alma Mater*, July 1899, p. 34.
56. Paul Crook, *Darwinism, War and History: The debate over the biology of war from the 'Origin of Species' to the First World War.* Cambridge University Press, Cambridge, 1994; Nancy Stepan, '"Nature's Pruning Hook": War, Race and Evolution, 1914–18', in *the Political Culture of Modern Britain*, (ed) J.M.W. Bean, London, 1987, pp. 129–48.
57. Frederic Wood Jones, 'Mechanism or Vitalism' in *Life and Living*, London, Kegan Paul, 1939, pp. 24–5.

CHAPTER THREE (PAGES 47–78)

1. *Speculum,* no. 25, July 1891, p. 77.
2. ibid., no. 1, July 1884, p. 4.
3. Bryan Gandevia, *The Melbourne Medical Students 1862–1942*, Melbourne Medical Students' Society, Melbourne, 1948, pp. 39–40.
4. *Speculum*, no. 9, May 1887, p. 14; *Alma Mater*, vol. vi, no. 5, May 1901, p. 143.
5. *Speculum*, no. 7, August 1886, p. 2.
6. Bryan Gandevia, op.cit., p. 16.
7. ibid., p. 41; *Alma Mater*, vol. 11, no. 4, July 1897, p. 11.
8. *Speculum*, no. 9, May 1887, p. 15.
9. ibid., no. 25, July 1891, p. 77.
10. For an interesting discussion about medical students and the working class patients see Janet McCalman, *Sex and Suffering,* pp. 171–94.
11. The relief of Mafeking was accomplished by the British forces in the Second Boer War in South Africa and made Robert Baden-Powell (later founder of the Scouting movement) a national hero. Australia sent troops to fight with the British.
12. *Speculum*, no. 47, September 1900, pp. 44–5.
13. ibid., no. 25, July 1891, p. 77. Clearly a play on 'soul' and 'sole', marching and the lyrics.
14. *Speculum,* no. 25, July 1891, p. 77.
15. Bryan Gandevia, *The Melbourne Medical Students 1862–1942*, frontispiece and p. 36.
16. *Speculum,* no. 2, December 1884, p. 6.
17. ibid.
18. ibid., no, 11, October, 1887, p. 40.
19. ibid., no. 39, November 1897, p. 74.
20. James C. Riley, *Rising Life Expectancy: A Global History*, Cambridge, Cambridge University Press, 2001, p. 38; *Speculum*, no. 8, February 1887, pp. 12–13.
21. J. E. Neild, 'Presidential Address', *Australasian Medical Gazette*, January 1901, p. 6.
22. *Speculum*, no. 62, May 1905, p. 2.

23. Roy Porter, *The Greatest Benefit to Mankind: A Medical History of Humanity from Antiquity to the Present*, Fontana, London, 1999, p. 680.
24. One of the first three students in the University of Melbourne Medical School was Patrick Moloney, who, as Medicus in *'On Massage', from The Box Seat, or, Medicus and the Shroff* (Melbourne: Melville and Mullen, 1900), author J. H. White (the Shroff), claimed university medicine had rediscovered the true and proper use of massage which had been perverted by alternative practitioners. Discussion on other therapies also indicates the paucity of the university *materia medica* of the time.
25. *Speculum*, no. 28, July 1892, p. 11.
26. Anne Hardy, *The Epidemic Streets: infectious disease and the rise of preventive medicine,* Oxford University Press, Clarendon Press, New York, 1993, p. 290.
27. *Speculum,* no. 23, February 1891, p. 145; *Speculum*, no. 24, April 1891, p. 3.
28. ibid., no. 29, September 1892, p. 79; 'A brilliant young Belgian brewer, Auguste de Bavay, met Pasteur on one of his trips and adopted his methods before emigrating to Melbourne, where he was employed at the Victoria Parade Brewery. Because of the warmer temperatures in Australia, de Bavay had to adapt Pasteur's methods to top fermentation. To protect against spoilage, he added greater amounts of hops, and fermented to a higher alcohol content. The highly fertile soils and plentiful sunshine resulted in increased levels of protein in the barley, which caused clouding of the beer, so de Bavay replaced some of the barley malt with cane sugar. The resulting beer was that now characteristic of Australia: light in colour and body, but tasting strongly of bitter hops. This style, although correctly described as a 'bitter' in Australia, is usually mistaken for a lager in Britain. Auguste de Bavay was quickly promoted to chief brewer at Fosters, a position he held from 1894 to 1904. For much of this time, he also acted as bacteriologist for the University of Melbourne and used the influence gained from that role to fight for improvements in the quality of the Melbourne water supply. In 1889 his denouncement of the city's fire plugs, claiming that they admitted sewerage and typhoid bacilli into the domestic water supply, resulted in a Royal Commission and the eventual removal of the devices. Although his primary interest was to protect the product of the brewery from infection, his efforts did much to protect the city's population likewise.' From 'The Beer of Revenge' Alan G Baxter 2000, a programme in the series 'Ockham's Razor' with Robyn Williams, broadcast 10/2/2002, Australian Broadcasting Commission in interview with Alan Baxter from the Centenary Institute in Sydney. His book, *Germ Warfare* which is about immunology is published by Allen & Unwin.
29. Roy Porter, op.cit., p. 681.
30. Joseph Lister (1827–1912) promoted the use of carbolic spray to disinfect instruments and wounds during surgery.
31. James Barrett, 'the Halford Oration', *Medical Journal of Australia*, 10 December 1932, p. 710.
32. Alan Gregory, *The Ever Open Door: A History of the Royal Melbourne Hospital*, Hyland House, South Melbourne, 1998, p. 166.

33. J. H. L. Cumpston, *The Health of the People: A study in federalism*, Roebuck, Canberra, 1978, p. 9.
34. T. S. Pensabane, op.cit., pp. 42–3; Warwick Anderson, op.cit., pp. 47–8.
35. *Argus*, 5 August 1892, p. 4.
36. Warwick Anderson, *The Cultivation of Whiteness*. See especially, part 1, The Temperate South.
37. Susan Lawrence, 'Medical Education' in Bynum, W.F., and Porter, Roy, *Companion Encyclopedia of the History of Medicine*, Routledge, London, 1997, vol. 2, p. 1164.
38. *Alma Mater*, vol iii, no. 6, September 1898, 'Brownless Memorial Supplement, being a Special Medical Supplement', p. 16.
39. ibid., pp. 27–8.
40. *Speculum*, no. 9, May 1887, p. 5.
41. *Alma Mater*, vol. 1, no. 5, May 1896, p. 15.
42. ibid.
43. ibid., vol. v, no. 3, June 1899, p. 53; 'stew' is slang for 'study'.
44. *Speculum*, no. 8, February 1887, pp. 1, 2.
45. ibid., no. 9, May 1887, p.1; Bryan Gandevia, op.cit., p. 15.
46. *Speculum*, no. 39, November 1897, p. 74.
47. ibid., no. 62, May 1905, p. 2.
48. 'The Profession in Victoria', *Medical Examiner,* 5 April 1877, p. 274, cited in T. S. Pensabane, op.cit., p. 58.
49. *Speculum*, no. 1, July 1884, p. 5.
50. ibid., no. 4, September 1885, p. 1.
51. ibid., no. 6, June 1886, p. 17.
52. ibid., no. 1, July 1884, p. 15.
53. Bryan Gandevia, op.cit., p. 16; *Speculum*, no. 1, July 1884, p. 5.
54. Bryan Gandevia, ibid., p. 15; H. B. Allen, *University of Melbourne, Medical School Jubilee,* p. 21. Among numerous examples see *Speculum*, no. 4, September 1885, p. 1; *Speculum*, no. 5, April 1886, p. 2; *Speculum*, no. 7, August 1886, p. 1.
55. *Speculum*, no. 3, May 1885, p. 21.
56. H. B. Allen, loc.cit.
57. *Speculum*, no. 7, August 1886, p. 2.
58. ibid., no. 1, July 1884, p. 5.
59. ibid., p. 6.
60. ibid., no. 5, April 1886, pp. 3–4.
61. ibid., no. 8, February 1887, p. 7; *Speculum*, no. 7, August 1886, p. 1. Ellery, Cutts and Brownless had also voted against the admission of women to the University and the Medical School. R. J. W. Selleck, *The Shop,* pp. 163, 270.
62. R. J. A. Berry, *Chance and Circumstance*, unpublished autobiography, University of Melbourne Medical Library, 1954, p. 109.
63. ibid., p. 110.
64. ibid.

65. ibid.
66. Harry Brookes Allen Papers, University of Melbourne Archives, Box 63/3, notebook on General Anatomy, Physiology and Pathology, vol. II, 1873.
67. H. B. Allen, op.cit., p. 9. There had been earlier dissections in Australia, though not in an educational institution. See H.P. Macdonald, *Human remains: episodes in human dissection*, Melbourne University Press, Carlton, 2005.
68. Harry B. Allen, op.cit., p. 13.
69. Alan Gregory, op.cit., p. 53.
70. Harry B. Allen, op.cit., p. 14. The appointment as pathologist launched Allen on a lengthy debate with the Council over his salary.
71. ibid., p. 18.
72. ibid., p. 19.
73. ibid., p. 20.
74. Alan Gregory, op.cit., pp. 165–6.
75. Harry B. Allen, op.cit., p. 23.
76. ibid., p. 24.
77. *Alma Mater*, vol iii, no. 6, September 1898, 'Brownless Memorial Supplement, being a Special Medical Supplement', p. 19.
78. ibid., p. 19.
79. ibid.
80. ibid.
81. ibid., pp. 27–8.
82. ibid.
83. *Speculum*, no. 40, May 1898, p. 26.
84. *University of Melbourne Calendar*, 1864–5, pp. 24–5, exemption for Anatomy '2. Notwithstanding anything here contained the Lectures in General Anatomy Physiology and Pathology and in Descriptive and Surgical Anatomy may be given and Dissections may be conducted during the months from May to October inclusive.' G.B. Halford in *Australian Medical Journal*, June 1873, p. 170.
85. Interview with James Guest, 3 June 2004.
86. *Speculum*, no. 62, May 1905, p. 2.
87. ibid.
88. R. J. W. Selleck, op.cit., p. 479.
89. *Speculum*, no. 63 August 1905, pp. 62–3. Tommy Bent was the ex land boomer and contemporary Premier of the State who had provided the money for the new Chair of Anatomy.
90. R. J. A. Berry, *Chance and Circumstance*, op.cit., p. 112.
91. *Alma Mater*, vol. 5, no. 3, June 1899, p. 63.
92. H. B. Allen, 'Earth and Sky', *Alma Mater*, vol. 6, no. 4, July 1899, p. 35.
93. *Speculum*, no 9, May 1887, pp. 5–6.
94. ibid., no. 3, May 1885, p. 11.
95. ibid., no. 7, August 1886, p. 1.
96. ibid., no. 16, January 1889, pp. 14–15.

97. ibid.
98. *Speculum*, no. 28 July 1892, p. 20.
99. ibid., no. 16, January 1889, pp. 14–15; *Speculum*, no. 41, July 1898, p. 30.
100. ibid.
101. R. J. A. Berry, *Chance and Circumstance*, op.cit., p. 112.
102. *Alma Mater*, vol iii, no. 6, September 1898, op.cit., p. 5.
103. Norcott to Allen, Nov. 22, 1893, H. B. Allen Papers, Box 15, University of Melbourne archives, University of Melbourne, Parkville.
104. ibid.
105. See H. B. Allen Papers, University of Melbourne archives.
106. R. J. A. Berry, op.cit., p. 110.
107. *Speculum*, October, 1887, no. 11, p. 12.
108. H. B. Allen, op.cit., p. 5.
109. *Alma Mater*, vol iii, no. 6, September 1898, op.cit., p. 25.
110. R. J. A. Berry, op.cit., p. 111.
111. K. F. Russell, op.cit., p. 105.
112. See H. B. Allen Papers, Box 63/b folder of correspondence, University of Melbourne archives.
113. *Speculum,* no. 2, December 1884, pp. 1–2. For discussion on this issue see R. J. W. Selleck, *The Shop,* op.cit., pp. 314–5; Laurence Geary, 'Australian Medical Students in 19th Century Scotland', *Proceedings of the Royal College of Physicians, Edinburgh*, vol. 26, 1996; 'The Scottish-Australian Connection, 1850–1900', in Vivian Nutton and Roy Porter (eds), *The History of Medical Education in Britain*, Rodpi, Amsterdam, 1995.
114. *Speculum,* no. 11, October 1887, p. 8.
115. ibid., no. 9, May 1887, p. 34.
116. '... their teaching system, especially anatomy, is excellent, but I don't think much of their hospital work.' *Speculum,* no. 10, July 1887, p. 24; *Speculum,* no. 37, May 1897, pp. 13, 14, 15.
117. *Speculum,* no 14, August 1888, pp. 29–30.
118. ibid., no. 10, July 1887, p. 10.
119. ibid., no. 20, March 1890, p. 1 (ed).
120. ibid.
121. *Speculum,* no. 28, July 1892, pp. 25–6; *Speculum*, no. 23, February 1891, p. 186.
122. ibid., no. 52, April 1902, p. 2.
123. ibid., p. 20.
124. T. S. Pensabane, op.cit., p. 58.

CHAPTER FOUR (PAGES 79–101)

1. F. Wood Jones, 'The legalization of the study of human anatomy' in J. V. Barry and A. E. Coates (eds), *The Proceedings of the Medico-Legal Society of Victoria during*

the years 1933–1934 and 1935–1936, vol. ii, Brown Prior Anderson Pty. Ltd, Melbourne, 1937, p. 135.

2. For a discussion of early dissections in the colonies see H. P. MacDonald, *Human Remains: episodes in human dissection*, Melbourne University Press, Carlton, 2005.
3. Based on the works of the Greek Hippocrates and the Roman Galen.
4. Ruth Richardson, *Death, Dissection and the Destitute*, Penguin, London, 1989.
5. Richard Evans, 'Epidemics and Revolutions', in Terence Ranger and Paul Slack (eds), *Epidemics and Ideas*, Cambridge University Press, Cambridge, 1993, p. 165; Roger French, 'The Anatomical Tradition' in Bynum, W. F., and Porter, Roy, *Companion Encyclopedia of the History of Medicine*, Routledge, London, 1997, vol. 1, pp. 100.
6. *Victorian Parliamentary Debates*, Legislative Council, 1862, vol. 8, p. 706.
7. ibid.
8. ibid., p. 837.
9. *Argus*, 7 August 1869, p. 4.
10. *Australian Medical Journal*, June 1873, p. 170.
11. ibid., pp. 170, 173.
12. *Argus*, 30 July 1869, p. 5.
13. ibid.
14. F. B. Smith, 'Disputes about Typhoid Fever in Victoria in the 1870s', op.cit., p. 2.
15. *Argus*, 30 July 1869, p. 5.
16. *Australian Medical Journal*, June 1873, pp. 170–71.
17. ibid.
18. *Argus*, 13 August 1869, p. 5.
19. *Australian Medical Journal*, June 1873, pp. 170, 172.
20. R. J. A. Berry, *Chance and Circumstance*, op.cit., p. 110.
21. *Argus*, 7 August 1869, p. 4.
22. *Argus Supplement*, 10 August 1869, p. 1.
23. *Australian Medical Journal*, June 1873, pp. 169–70.
24. ibid., p. 171.
25. *Argus*, 7 August 1869, p. 4.
26. H. B. Allen, *University of Melbourne, Medical School Jubilee 1914*, op.cit., p. 107.
27. ibid., pp. 107–8.
28. Halford to University Council, University of Melbourne Registrar (UM 419) 93/46 Correspondence in Medical School Folder 3 (1869–1909).
29. ibid.
30. See Cadaver Archives, Department of Anatomy and Cell Biology, University of Melbourne, Parkville.
31. *Speculum*, no. 28, July 1892, pp. 20–1.
32. R. J. W. Selleck, op.cit., p. 242.
33. *Speculum*, no. 11, October 1887, p. 12.
34. ibid.
35. ibid., p. 14.
36. ibid., p. 13.

37. ibid.
38. ibid.
39. ibid., p. 47.
40. ibid.
41. ibid.
42. ibid., no 12 February 1888, p. 13.
43. R. J. W. Selleck, loc.cit., p. 242.
44. *Speculum*, no. 16, January 1889, p. 1.
45. ibid., pp. 17–8.
46. ibid.
47. ibid., pp. 13–4, 17–18.
48. *Speculum*, no. 17, June 1889, pp. 13–4.
49. ibid., no. 22, September 1890, p. 142.
50. ibid., no. 25, July 1891, p. 51.
51. ibid., no. 30, August 1894, p. 30.
52. ibid., no. 31, May 1895, p. 31; *Speculum*, no. 38, July 1897, p. 72; *Speculum*, no. 39, November 1897, p. 74.
53. *Speculum*, no. 40, May 1898, p. 26.
54. ibid., p. 27.
55. K. F. Russell, *The Melbourne Medical School, 1862–1962*, op.cit., pp. 74f. There is a great need for more detailed work to be published on the women medical students.
56. *Speculum*, no. 11, October 1887, p. 40.
57. ibid., no. 21, June 1890, p. 95.
58. ibid.
59. Bryan Gandevia, The Melbourne Medical Students, 1862–1942, op.cit., p. 47.
60. *Alma Mater*, vol. 5, no. 3, June 1900, p. 108.
61. *Speculum*, no. 21, June 1890, p. 66.
62. ibid.
63. ibid.
64. ibid.
65. ibid., p. 67.
66. ibid.
67. ibid, pp. 67–8.
68. ibid.
69. ibid.
70. R. J. A. Berry, loc.cit., p. 110.
71. Cadaver Archives held in the Department of Anatomy and Cell Biology, The University of Melbourne. First receipt in small book in envelope.
72. ibid.
73. There were forty-seven male and only eight female corpses.
74. *Speculum*, no. 30, August 1894, p. 12.
75. See Cadaver Archives op. cit.
76. *Speculum*, no. 2, December 1884, p. 7

77. ibid., no. 25, July 1891, pp. 78–9.
78. ibid.
79. ibid.
80. *Alma Mater*, vol. iv, no. 1, April 1899, p. 19.
81. See Springthorpe diaries in the Manuscripts collection, State Library of Victoria.
82. This is a well-attested psychological reaction. See the work of Raymond Gibbs, Herbert Clark and, of course, Sigmund Freud.
83. *Speculum*, no. 28, July 1892, pp. 20–1; *Speculum*, no. 18, August 1889, p. 47.
84. ibid., no. 6, June 1886, p. 1, ed. 'only a small proportion of our members are able to establish themselves in a large town ... the greater number settle in outlying districts in the country.'
85. *Speculum*, no. 10, July 1887, p. 10.
86. ibid., no. 9, May 1887, p. 1.
87. ibid., no 13, May 1888, p. 6.
88. ibid., no. 29, September 1892, p. 77.
89. In the early nineteenth century in Britain, the study of anatomy was a point of political dispute and dissection was a tool of radicals attempting to upset the received order underpinned by creationism. See Adrian Desmond, *The Politics of Evolution: Morphology, Medicine, and Reform in Radical London*, University of Chicago Press, Chicago, 1989.

CHAPTER FIVE (PAGES 103–131)

1. Edward Dyson, 'Scientific Selection', *Lone Hand*, 1907, p. 161.
2. *Argus*, 1 December 1917, p. 19; *Age*, 1 December 1917, pp. 13–14.
3. *Argus*, 1 December 1917, p. 19.
4. ibid.
5. Conscription was supported by Melbourne's Protestant middle classes but strongly opposed by both the Roman Catholic Church, led by Archbishop Daniel Mannix, as well as many on the left of politics. Ultimately, this referendum was to suffer the same fate as its predecessor in 1916, being narrowly rejected at the ballot box.
6. R. J. A. Berry, *Chance and Circumstance*, op.cit.
7. ibid., p. 24.
8. ibid., pp. 24–5.
9. ibid., p. 34.
10. *The Australasian Students' Song Book*, published for the Song Book Board by George Robertson, Melbourne, 1911, p. 18.
11. *Herald*, 9 July 1923.
12. Ronald Ridley (ed), *The Diary of a Vice-Chancellor, University of Melbourne, 1935–1938*, Raymond Priestley, Melbourne University Press, Carlton South, 2002, p. 34.
13. R. J. A. Berry, op.cit., p. 111.
14. Sydney Sunderland, 'The Melbourne Medical School and some of its "characters" 1931–1975', *Chiron: Journal of the University of Melbourne Medical Society*, vol. 2, no. 2, 1992, p. 46.

15. *Speculum*, no. 124, May 1929, p. 86.
16. Sydney Sunderland, op.cit., pp. 46–7.
17. ibid.
18. Characteristically, Barrett attempted to rewrite history on this point. See James Barrett, 'The Halford Oration', *Medical Journal of Australia*, 10 December 1932, pp. 708f.
19. Ronald Ridley, loc.cit.
20. Frederic Wood Jones to Arthur Keith, 10 April 1929, Arthur Keith Papers, Manuscript 174 Correspondence with Frederick Wood Jones (1905–1951), The Royal College of Surgeons of England, London.
21. R. J. A. Berry, op.cit., p. 115.
22. ibid., pp. 189–90.
23. R. J. A. Berry and R. G. Gordon, *The Mental Defective: A Problem in Social Inefficiency*, Kegan Paul, London, 1931; Kevin Blackburn, 'Preaching The Gospel of Efficiency: The promotion of Ideas about Profit-sharing and Payment by Results in Australia, 1915–1929', *Australian Historical Studies*, vol. 27, no. 107, October 1996, pp. 257–280; E. J. T. Brennan (ed), *Education for National Efficiency: the Contribution of Sidney and Beatrice Webb*, Athlone Press, London, 1975; G. R. Searle, *The Quest for National Efficiency: A Study in British Politics and Political Thought, 1899–1914*, Ashfield Press, London, 1990 (first edition 1971).
24. I have mentioned the literature on this in chapter two.
25. R. J. A. Berry and S. D. Porteus, 'Size of Brain and Intelligence', *Medical Journal of Australia*, 18 August 1917, p. 155.
26. Warwick Anderson, *The Cultivation of Whiteness*, op.cit., pp. 67f.
27. For example the *Herald*, 26 October 1916.
28. See Michael Roe, *Nine Australian Progressives: Vitalism in Bourgeois Social Thought 1890–1960*, University of Queensland Press, St. Lucia, 1984.
29. *Herald*, evening, 9 May 1929.
30. ibid.
31. *Victorian Parliamentary Debates (VPD)*, vol. 179, 13 August 1929, pp. 806, 815.
32. *Medical Journal of Australia*, 26 July 1919, p. 72.
33. *Sun News-Pictorial*, 11 July 1923.
34. *Age*, 3 August 1928, p. 8.
35. ibid.
36. R. J. A. Berry, *Report to The Edward Wilson (of 'The Argus') Trust on Mental Deficiency in the State of Victoria, with Suggestions for the Establishment of a Child Guidance Clinic*, Argus, Wilson and Mackinnon, Melbourne, 1929, p. 6.
37. K. F. Russell, *The Melbourne Medical School*, op.cit., p. 157.
38. *VPD*, vol. 172, 6 October 1926, p. 1855.
39. 'Observations (mainly psychological) on the Concept of Mental Deficiency', *Medical Journal of Australia*, 7 February 1925, p. 136. For a biography of Miller see Michael Roe, *Nine Australian Progressives*, op.cit., pp. 280–315. Miller was initially educated in Victoria and returned to speak about mental deficiency in the 1930s.

40. *VPD*, vol. 172, 6 October 1926, p. 1853, and vol. 179, 13 August 1929, pp. 807–8.
41. ibid. vol. 179, 13 August 1929, p. 801.
42. ibid. vol. 172, 6 October 1926, p. 1854.
43. R. J. A. Berry, *Chance and Circumstance*, op.cit., p. 143.
44. *Medical Journal of Australia*, 8 August 1925, pp. 180–1.
45. *VPD*, vol. 179, 13 August 1929, pp. 803, 805.
46. ibid. William Ernest Jones, *Report on Mental Deficiency in the Commonwealth of Australia*, H.J. Green, Government Printer, Canberra, 1929, p. 7.
47. *Argus*, 14 December 1926.
48. Berry wrote that '[t]he truth is, of course, that about one in every ten of the population whom you meet in your daily pursuits has the physical and sexual passions of the adult with the brain of an under-developed child'. See Berry, *Chance and Circumstance*, p. 143.
49. *Argus*, 5 September 1929.
50. 'Sterilization of the Unfit', *Medical Journal of Australia*, 21 November 1931, p. 655.
51. *Western Daily Press*, 12 July 1934.
52. *Argus*, 5 July 1923.
53. *VPD*, vol. 208, 31 October 1939, p. 1838.
54. Kenan Malik, *The Meaning of Race: Race, History and Culture in Western Society*, New York University Press, New York, 1996, p. 119.
55. See S. J. Gould, *The Mismeasurement of Man*, Norton, New York, 1981.
56. Stanley Porteus, *A Psychologist of Sorts: The Autobiography and Publications of the Porteus Maze Tests*, Pacific Books, Palo Alto, California, 1969, p. 56.
57. R.J.A. Berry, 'The Problem of the Unfit', *Herald*, 3 May 1924, p. 11.
58. *Ballarat Star*, 2 October 1912. Richard J. A. Berry and A. W. D. Robertson, *The place in nature of the Tasmanian Aboriginal as deduced from a study of his calvaria, Part 2, His relation to the Australian Aboriginal*, Edinburgh, Robert Grant & Son, 1914. Reprint from *Proceedings*, the Royal Society of Edinburgh. Session 1913–1914, vol. 34, part 2 (no. 12).
59. R.J.A. Berry, *Chance and Circumstance*, p. 132. A large proportion of Berry's research work with Porteus was an attempt to refine Pearson's work on the correlation between head size and intelligence, *Medical Journal of Australia*, 23 June, 1917, pp. 532–544; Berry, op.cit., p. 134.
60. R.J.A. Berry's archive *DDX 76*. Original documents held by the Lancashire Record Office, Preston, Lancashire, England, undated newspaper clipping.
61. ibid. See place card and signature.
62. Daniel Kevles, *In the Name of Eugenics*, Harvard University Press, London, England, 2001 (1985), pp. 32–3.
63. ibid., p. 77.
64. ibid., p. 20.
65. 'Report of Psychical Research Committee', *Report of the 6th Meeting of the Australasian Association for the Advancement of Science, Brisbane 1895*, Brisbane, 1896, pp. 842–3.

66. ibid.
67. Daniel Kevles, op.cit., p. ix.
68. *Speculum*, no. 75, October 1909, p. 103.
69. R.J.A. Berry, 'Physical and Mental Testing', *Medical Journal of Australia*, 12 September 1925, p. 352.
70. ibid., p. 354.
71. ibid., p. 355.
72. 'Though it could not be too strongly urged that diagnoses of mental inefficiency could not be based (as was far too frequently done) on Binet-Simon tests alone', ibid. Berry's methodology was utilised by a number of doctors: see Drs A. P. Derham and Geoffrey A. Pennington, *Medical Journal of Australia*, 15 January 1927, p. 101.
73. *Medical Journal of Australia*, 14 December, 1918, p. 489.
74. ibid.
75. ibid.
76. ibid.
77. R. J. A. Berry, *Chance and Circumstance*, op.cit., p. 188.
78. For some examples of the numerous sources relating to the extensive and successful campaign conducted by Berry see R.J.A. Berry, *Report to the Edward Wilson (of 'The Argus') Trust on Mental Deficiency in the State of Victoria, with Suggestions for the Establishment of a Child Guidance Clinic*, Wilson and Mackinnon, Melbourne, 1929, pp. 4–5; John Lewis, Removing the Grit: The Development of Special Education in Victoria 1887–1947, unpublished PhD thesis, LaTrobe University, 1989, pp. 200–1 for 'Professor Berry's Campaign'.
79. *Argus*, 14 December 1926; 15 December 1926; 16 December 1926.
80. *Argus*, 14 December 1926.
81. ibid.
82. *Herald*, 4 July 1917.
83. ibid.
84. ibid.
85. Berry, *Chance and Circumstance*, p. 143. For a list of Berry's extensive travelling and lecturing see R. J. A. Berry, *Report to the Edward Wilson (of 'The Argus') Trust*.
86. *Herald*, 6 July 1917.
87. ibid., Don Garden, *Theodore Fink: A Talent for Ubiquity*, Melbourne University Press, Carlton, 1998, p. 121.
88. *Herald*, 6 July 1917. Also, to add further educational significance to the occasion, Berry's and Porteus' lecture had been organised by the Council of Public Education, the body set up to advise the Minister of Education.
89. ibid.
90. *Medical Journal of Australia*, 23 June 1917, p. 536. Tate was also present at this meeting. He was singled out for a special welcome from those present at the seminar by the chairman who announced that 'without his co-operation

and assistance, the results of the work to be submitted to them by Professor Berry could not have been carried out'. The chairman was referring to the financial assistance Tate had supplied to Berry in his official role by providing him with two assistants from the Education Department: a Miss Temple and Stanley Porteus, the latter having been seconded part-time from his duties as headmaster of the Bell Street Special School to work in Berry's Department. As well, Tate gave Berry the opportunity to measure the heads of many thousands of State school students. See Berry, *Report to the Edward Wilson (of 'The Argus') Trust*, p. 3. Tate's association with Porteus continued for many years. See R. J. W. Selleck, *Frank Tate: A Biography*, Melbourne University Press, 1982, p. 273.

91. ibid.
92. ibid., p. 537.
93. ibid., p. 536.
94. ibid., p. 539.
95. R.J.A. Berry, *Chance and Circumstance*, p. 115.
96. *Medical Journal of Australia*, 23 June 1917, p. 540.
97. R.J.A. Berry, op.cit., p. 135.
98. *Medical Journal of Australia*, 23 June 1917, pp. 540–1.
99. ibid., p. 541.
100. For an extended discussion of these Bills see Ross L. Jones, 'The Master Potter and the Rejected Pots: Eugenic Legislation in Victoria 1918–1939', *Australian Historical Studies*, vol. 30, no. 113, October 1999, pp. 319–342.
101. Tasman Lovell, 'The Tasmanian Mental Deficiency Act' *The Australasian Journal of Psychology and Philosophy*, vol. 1, 1923, pp. 285–289.
102. For a summary of this legislative programme, see John Lewis, Removing the Grit: The Development of Special Education in Victoria, 1887–1947, PhD thesis, Latrobe University, 1989, pp. 267–70; John Lewis, 'So Much Grit in the Hub of the Educational Machine: Schools, Society and the Invention of Measurable Intelligence' in Bob Bessant (ed) *Mother State and Her Little Ones: Children and Youth in Australia 1860s–1930s*, Centre for Youth and Community Studies, Melbourne, 1987, p. 161; Stephen Garton, 'Sound Minds and Healthy Bodies: Re-considering Eugenics in Australia, 1914–1940', *Australian Historical Studies*, vol. 26, no. 103, October 1994, p. 164.
103. For a description of Berry as an authoritarian, conservative social Darwinist, see Mary Cawte, 'Craniometry and Eugenics in Australia: R.J.A. Berry and the quest for Social Efficiency', *Historical Studies*, vol. 22, no. 86, April 1986, pp. 35–53, contra Kenneth F. Russell, 'Richard James Arthur Berry, 1867–1962', in Geoffrey Kenny and Harold Attwood (eds) *Festschrift for Kenneth Fitzpatrick Russell*, Queensberry Hill Press, Melbourne, 1978, pp. 25–45.
104. Berry spoke against the Carnegie endowment at Edinburgh University 'in so far as it sought to make University education free, remains to this day, a colossal and stupendous failure'. R. J .A. Berry, 'The Modern University and its Upkeep' in Henry Payne (ed) *Proceedings of the First Educational Congress*, State of Victoria, held in

Melbourne, 1912, Mullet, Melbourne, [no date], p. 129. There had been an earlier attempt to abolish fees at the University of Melbourne. In 1891 a private member's Bill in the Legislative Assembly proposing a free university in the legislative assembly was defeated. See John Rickard, *H.B. Higgins: the Rebel as Judge*, George Allen & Unwin, Sydney, 1984, p. 155.

105. See Ross L. Jones, Skeletons in Toorak and Collingwood cupboards: Eugenics in educational and health policy in Victoria, 1910–1939, unpublished PhD thesis, Monash University, 2000.
106. *Argus*, 6 November 1912. The research was published as Richard J. A. Berry and L. W. G. Büchner, 'The Correlation of Size of Head and Intelligence as Estimated from the Cubic Capacity of Brain of 355 Melbourne Criminals' [read 10 October, 1912], *Proceedings of the Royal Society of Victoria*, 25 (N.S.), Pt. II., 1913, pp. 229–53.
107. *Argus*, 6 November 1912.
108. ibid., 9 November 1912.
109. ibid.
110. ibid.
111. ibid.
112. ibid.
113. *Argus*, 11 November 1912.
114. ibid.
115. ibid.
116. Sydney Sunderland, 'The Melbourne Medical School and some of its "characters" 1931–1975', *Chiron: Journal of the University of Melbourne Medical Society*, vol. 2, no. 2, 1992, p. 48.
117. *The Australasian Students' Song Book*, published for the Song Book Board by George Robertson, Melbourne, 1911, p. 18. Roentgen rays were later renamed X-rays.
118. Review of Richard Berry's *Brain and Mind* by T. R. A. 'Professor Berry Examines the Head-End of Man', *Herald*, 7 July 1928.
119. Sydney Sunderland, loc.cit.
120. R. J. A. Berry, *Chance and Circumstance*, pp. 187–90.
121. Penrose, Memoirs—1964: Phenylketonuria, Lionel S. Penrose Papers, University College London Archives, quoted in Kevles, *In the Name of Eugenics*, p. 159.
122. For the decline of Berry's views in Britain see Elazar Barkan, *The Retreat of Scientific Racism: Changing Concepts of Race in Britain and the United States between the two World Wars*, Cambridge University Press, Cambridge, 1992.
123. *Farrago*, vol. 1, no. 2, April 17, 1925, p. 1.
124. ibid.
125. Goddard, Henry Herbert, *The Kallikak Family: A Study in the Heredity of Feeble-mindedness*, Macmillan, New York, 1921.
126. *Farrago*, loc.cit.
127. See also 'Call to rename uni building' *Sunday Age*, 23 February, 2003.
128. The classic analysis of changes in scientific fashion is in Thomas S. Kuhn, *The*

Structure of Scientific Revolutions, University of Chicago Press, Chicago, 1996 (1967).

CHAPTER SIX (PAGES 133–161)

1. Frederic Wood Jones, 'The Changing Point of View', Joseph Bancroft Memorial Lecture, Brisbane, June 5 1931, *Life and Living*, Kegan Paul, London, 1939, p. 160.
2. Frederic Wood Jones, 'Mechanism or Vitalism?', Livingstone lecture no. 1, Sydney, 1935, *Life and Living*, Kegan Paul, London, 1939, p.19.
3. Frederic Wood Jones, 'The Spirit of Adventure' Medical Congress, Adelaide, 5 August, 1937 in *Life and Living*, Kegan Paul, London, 1939, p. 45.
4. Frederic Wood Jones, *Buchanan's Anatomy*, Bailliere, Tindeall and Cox, London, 1943; *Structure and Function as Seen in the Foot*, Bailliere, Tindeall and Cox, London, 1944; *Hallmarks of Mankind*, Bailliere, Tindeall and Cox, London, 1948.
5. Frederic Wood Jones, *Design and Purpose*, Kegan Paul, London 1942; *Habit and Heritage*, Kegan Paul, London 1943; and *Trends of Life*, Edward Arnold, London, 1953.
6. B. E. Christophers, 'Frederic Wood Jones: Coral and Atolls', *Australian and New Zealand Journal of Surgery*, no. 65, 1995, pp. 752. Using the middle name was an affectation taken on by numerous anthropologists and medical men of the era. For a bibliography of Wood Jones's publications see the numerous compilations of B. E. Christophers.
7. Wilfrid E. Le Gros Clark, 'Frederic Wood Jones 1879–1954', *Biographical Memoirs of Fellows of the Royal Society*, vol. 1, November 1955, p. 119.
8. A. Keith, *An Autobiography*, Watts, London, 1950, p. 273. Frank Spencer (ed), *History of Physical Anthropology*, Garland, NewYork, 1997, p. 553.
9. So much had been written on Piltdown man, and there are so many different schools of thought, it is impossible to give a definitive or short list. Try Frank Spencer, *Piltdown: a scientific forgery*, Oxford University Press, London, 1990.
10. B. E. Christophers, 'Frederic Wood Jones: His Illnesses', *Australian and New Zealand Journal of Surgery*, no. 70, 2000, p. 526.
11. Wood Jones to Mrs Keith, 28 May 1905, Singapore. Correspondence between Sir Arthur Keith and Frederic Wood Jones, 1905–1951, p. 2, Arthur Keith Papers, The Royal College of Surgeons of England Archives, MSS Add 174, London.
12. Wood Jones to Mrs Keith, ibid., back of p. 1, The Royal College of Surgeons of England Archives, MSS Add 174, London. For his lifelong asceticism see A. Keith, *An Autobiography*, Watts, London, 1950, p. 657. It is thought that he was without the sense of smell (anosmic). Interview with James Guest, 3 June 2004.
13. Wood Jones to Keith, Cocos Keeling Islands, Indian Ocean, September 1905. Correspondence between Sir Arthur Keith and Frederic Wood Jones, 1905–1951, p. 4 front and back, The Royal College of Surgeons of England Archives, MSS Add 174, London.
14. Francis Darwin (ed), *Selected Letters on Evolution and Origin of Species*, Dover, NY, 1952 (1898), p. 176.

15. William Cowper, *The Task*, 1785.
16. Wood Jones to Keith, Cocos Keeling Islands, op.cit., For a detailed discussion see B. E. Christophers, 'Frederic Wood Jones: Coral and Atolls', *Australian and New Zealand Journal of Surgery*, no. 65, 1995, pp. 749–60.
17. ibid.
18. C. W. Andrews to Wood Jones, 27 May 1909, headed: British Museum (Natural History) Cromwell Road, London SW. Frederic Wood Jones, Other Correspondence, The Royal College of Surgeons of England Archives, MSS 344 and 546, London.
19. Including a review in *Nature* by Stanley Gardiner, *Nature*, 1910; 84, p. 432–3.
20. B. E. Christophers, 'Frederic Wood Jones: His major books and how they were reviewed', *Australian and New Zealand Journal of Surgery*, no. 67, 1997, p. 646.
21. Wilfrid E. Le Gros Clark, 'Frederic Wood Jones. 1879–1954', *Biographical Memoirs of Fellows of the Royal Society*, vol. 1, November 1955, p. 120.
22. ibid., pp. 120–1.
23. Initially in 1963 the term 'seagull' was used, and replaced later by the more poetic 'butterfly'.
24. *The Mistress of Vision*, Poems (1913) in Frederic Wood Jones, 'The Changing Point of View', Joseph Bancroft Memorial Lecture, Brisbane, p. 214.
25. ibid.
26. Frederic Wood Jones, *Life and Living*, p. 115.
27. ibid., p. 114.
28. Davidson Black to Wood Jones 28 March 1923. Frederic Wood Jones, Other Correspondence, The Royal College of Surgeons of England Archives, MSS 344 and 546, London. Peking Man is sometimes now called Beijing Man, also Sinanthropus pekinensis (currently Homo erectus pekinensis).
29. Frederic Wood Jones, 'Anatomy and a Life Principle' Commemoration address, Adelaide University, 1923, *Life and Living*, p. 120.
30. There are still only a few hundred residents on the atoll due to its remoteness: 12° 09' south of the equator and 96° 53' east of Greenwich. The first airstrip was built in 1997.
31. B. E. Christophers, 'Frederic Wood Jones: Coral and Atolls', p. 753.
32. Throughout the nineteenth century in Britain there occurred a new 'scientific' definition of race, particularly as a consequence of Darwin's rise in influence. I have discussed this in the previous chapter. The consequence of this new idea of race was a radical realignment of the attitudes of Europeans to non-Europeans, with some striking manifestations. For example, in British India 'going native' and marrying a 'coloured' bride of a similar social status was considered acceptable in the eighteenth century, but became more and more taboo in the nineteenth century. See William Dalrymple, *White Mughals: Love and Betrayal in Eighteenth-Century India*, Viking, London, 2003; V. G. Kiernan, *The Lords of Human Kind: black man, yellow man, and white man in an age of empire*, London, Cresset Library, 1988, c.1969.
33. Wood Jones to Keith, op.cit., p. 5 back, The Royal College of Surgeons of England Archives, MSS Add 174, London.

34. This was a humorous quip for the benefit of Wood Jones by the celebrant and friend Hugh Chapman. See B. E. Christophers, 'Frederick Wood Jones and Atolls', p. 759.
35. ibid., pp. 758–9.
36. A. P. Elkin (ed), *Grafton Elliot Smith: the Man and his Work*, Sydney University Press, Sydney, 1974, p. 30.
37. Interview with James Guest, 3 June 2004.
38. Frederic Wood Jones to Arthur Keith, 9 November 1936. Correspondence between Sir Arthur Keith and Frederic Wood Jones, 1905–1951, p. 172–3, The Royal College of Surgeons of England Archives, MSS Add 174, London. The full text reads: 'I am sincerely sorry for poor old E.S. He seems to have come to a very sad old age. His wife, his sons – everything. I fancy that at the bottom of all his troubles is the outcome of the fact that he is an Australian, and that he married an Australian wife. I don't expect you will appreciate this in the way I do – for I have lived with Australians since 1919. His wife was a brilliant violinist, a charming woman – a Sydney girl, no soul, but in the place of soul, material social ambition. He – from Grafton – had a good brain, great charm, and a lot of Huxley's characteristics. But – here you can laugh at a man named Jones, born in Hackney – he was at bottom, like so many Australians, lacking in those things, loyalty, sincerity, altruism – and the hundreds of other unspoken things, that constitute the basis of (hated word) a gentleman. I would have been very loyal to E.S., for I learned to be very fond of him in Nubia – when he was at his best: away from rivals, away from controversies, away from jealousies – and in camp – he was the finest companion one could wish for. But social ambition, mainly due to his wife – intense jealousy of his fellows – intolerance – and a little devious streak, not altogether nice, made him hard to serve as a loved master. Still I shall always owe him a debt, for the days in camp in Nubia – and I am very sorry his old age is being a sad one. As a matter of fact, I vastly prefer him to J. P. Hill – a complaining little creature with a shrivelled soul – or so he always seemed to me. I never liked a man who whines – and J.P. always whines.' J. P. Hill was one of the Sydney triumvirate that achieved fame in anatomy and anthropology in the twentieth century – the other two being Elliot Smith and J. T. Wilson – see Patricia Morison, *J. T. Wilson and the Fraternity of Duckmaloi*, Atlanta, Amsterdam, 1997.
39. Warren R. Dawson (ed), *Sir Grafton Elliot Smith: a biographical record: by his colleagues*, Cape, London, 1938. Elliot Smith received a large advance for his autobiography from Jonathon Cape. On many evenings his socialite wife would entertain and he would retire early claiming he needed to work on his autobiography. On his death it was discovered that he had written nothing. The publishers demanded their advance back and his old colleagues and friends agreed to write the above work to placate the publishers.
40. Frederic Wood Jones, 'The Women's School to War', numbered 51, 25 pages of typescript in unpublished autobiographical notes. Letters to Dr Ulrica Hubbe, 1927–1954, The Royal College of Surgeons of England Archives, MSS 344 and 546, London.
41. ibid.

42. Ronald Ridley (ed), *The Diary of a Vice-Chancellor,* p. 16. Priestley was also very nervous of Wood Jones's presence when he lectured himself. See ibid., p. 31.
43. Interview with James Guest, 3 June 2004.
44. Ronald Ridley, op.cit., p. 407.
45. Wilfrid E. Le Gros Clark, 'Frederic Wood Jones, 1879–1954, p. 122.
46. ibid.
47. Sydney Sunderland, 'The Melbourne Medical School and some of its "characters" 1931–1975', p. 52.
48. Interview with James Guest, 3 June 2004. I will discuss the McCoy Society in the next chapter.
49. G. Eliott Smith to Wood Jones, 12 December 1923. Frederic Wood Jones, Other Correspondence, The Royal College of Surgeons Archives, MSS 344 and 546, London.
50. Chris Beard, *The Hunt for the Dawn Monkey: Unearthing the Origins of Monkeys, Apes and Humans*, University of California Press, Berkeley, 2004, p. 122.
51. G. Elliot Smith, *Proceedings of the Zoological Society of London*, 1919, Longmans, London, 1919, p. 470.
52. ibid., p. 472.
53. P. Chalmers Mitchell, *Proceedings of the Zoological Society of London*, 1919, Longmans, London, 1919, p. 496.
54. Professor McBride, *Proceedings of the Zoological Society of London*, 1919, Longmans, London, 1919, pp. 497, 498.
55. Chris Beard, *The Hunt for the Dawn Monkey*, op.cit., p. 280.
56. Frederic Wood Jones, *The Problem of Man's Ancestry*, SPCK, London, 1918.
57. See, for example, H. S. Harrison's review, 'Man's Place among the Mammals', Edward Arnold, London, 1929', *Man*, nos. 31–33, February 1931, p. 30.
58. A. S. Romer's review, 'Man's Place among the Anthropoids', *American Anthropologist*, vol. 39, no. 4, part 1 (Oct.—Dec., 1937), p. 703.
59. Ashley Montagu was a maverick like Wood Jones. All their academic encounters were, strangely enough, hostile, sharing, as they did, many similar views. Only Darwinism separated them. 'UNESCO man' was the statement of culture and race issued by the United Nations after the exposure of the Nazi race plans after 1945. It was a direct rebuttal of the biological view of race that had begun with Darwin and ended with the Holocaust and the eugenics movement. This instituted a period when biological determinism was unfashionable, especially in the social sciences in the 1950s and 60s. Of eugenics in 1965, the American social scientist, Donald T. Campbell, wrote 'the rejection was overwhelming, and persists in strength today' in 'Variation and Selective Retention in Socio-cultural Evolution', in Barringer, Herbert R. et al (eds), *Social Change in Developing Areas: A Reinterpretation of Evolutionary Theory*, Schenkman, Cambridge, Massachusetts, 1965, p. 23.
60. Francis M. Ashley Montagu, 'The Tarsian Hypothesis and the Descent of Man', *The Journal of the Royal Anthropological Society of Great Britain and Ireland*, vol. 60 (July to December 1930), pp. 335–62.
61. ibid., p. 335.

62. ibid., p. 362. He fought another anatomical battle with Montagu in the 1950s; the last of a series being in *Man*, vol. 49 (March 1949), p. 35.
63. For the anatomical details that support the theory see Chris Beard, *The Hunt for the Dawn Monkey*, p. 133–5. For modern adherents see pp. 126–41.
64. Frederic Wood Jones to Arthur Keith, 24 November 1929. Correspondence between Sir Arthur Keith and Frederic Wood Jones, 1905–1951, no. 89, The Royal College of Surgeons of England Archives, MSS Add 174, London. This almost certainly applies to the meeting in 1919 of the Zoological Society in which Elliot Smith deserted Wood Jones's Tarsian Hypothesis (without warning it seems).
65. Frederic Wood Jones, 'The Women's School to War', numbered 51, 25 pages of typescript in unpublished autobiographical notes. Letters to Dr Ulrica Hubbe, 1927–1954, pp. 13–15; 18; 21–27, The Royal College of Surgeons of England Archives, MSS 344 and 546.
66. Arthur Keith to Frederic Wood Jones, 27 February 1920, op.cit. At this time Keith had not read the book.
67. I have discussed the late Victorian revulsion with natural selection in chapter two.
68. Frederic Wood Jones, *Habit and Heritage*, pp. 10–11. See also p. 11 for his indictment of Tennyson's phrase 'nature red in tooth and claw', p. 11.
69. Frederic Wood Jones, 'The Women's School to War', p. 24.
70. J. M. T. Carter, *Painting the Islands Vermilion: Archibald Watson and the Brig Carl*, Melbourne University Press, Carlton South, 1999. A genuine eccentric, Watson's dark secret (known to Wood Jones) was an early career as a blackbirder (slaver) in the Pacific.
71. Egon Erwin Kisch, *Australian Landfall*, translated by John Fisher and Irene and Kevin Fitzgerald, Australasian Book Society, Sydney, 1969 (1937), p. 209. Kisch, a friend of Franz Kafka and a renowned reporter credited with inventing the style of reporting known as 'reportage', had been invited to visit Australia to speak against Hitler by the World Committee Against War and Fascism. After being refused permission to land at Fremantle, he leapt ashore at Melbourne, breaking his leg. He prolonged his stay through the means of legal proceedings to the High Court. Justice Evatt, later Leader of the Labor Party, threw out the plea against Kisch of 'subversive activities', resulting in Kisch being given the famous dictation test in Scottish Gaelic, administered as a consequence of his well-known fluency in most European languages. He later wrote that many Australian Labor leaders avoided fighting for a general improvement of conditions for all workers and rather settled on attacking the point of least resistance, and 'the point of least resistance was the Chinese'.
72. Interview with James Guest, 3 June 2004.
73. Richard Berry to Frederic Wood Jones, 6 February 1920. Frederic Wood Jones, Other Correspondence, The Royal College of Surgeons of England Archives, MSS 344 and 546.
74. ibid.
75. Frederic Wood Jones, South Australia I: numbered 53 in handwriting, 25 typewritten pages in unpublished autobiographical notes, Deed Box, Letters and Autobiography

(unnumbered), The Royal College of Surgeons of England Archives, London.

76. Frederic Wood Jones to Arthur Keith, 24 August 1927. Correspondence between Sir Arthur Keith and Frederic Wood Jones, 1905–1951, no. 69, The Royal College of Surgeons of England Archives, MSS Add 174, London.
77. For a general discussion see Peter J. Bowler, *Evolution: the History of an Idea*, University of California Press, Berkeley, 1989, p. 333f. For some modern examples see Edward J. Steele, Robyn A. Lindley, Robert V. Blanden, *Lamarck's Signature: How Retrogenes are Changing Darwin's Natural Selection Paradigm*, Perseus Books, 1998; Eva Jablonka and Marion J. Lamb, *Epigenetic Inheritance and Evolution: The Lamarckian Dimension*, Oxford University Press, 1995; C. H. Waddington, *The Evolution of an Evolutionist*, Cornell University Press, 1975.
78. Frederic Wood Jones, *Habit and Heritage*, p. 14.
79. Frederic Wood Jones to Arthur Keith, 10 December 1926. Correspondence between Sir Arthur Keith and Frederic Wood Jones, 1905–1951, no. 63–4, The Royal College of Surgeons of England Archives, MSS Add 174, London.
80. ibid.
81. ibid.
82. Frederic Wood Jones to Arthur Keith, 5 October 1926. Correspondence between Sir Arthur Keith and Frederic Wood Jones, 1905–1951, p. 54, The Royal College of Surgeons of England Archives, MSS Add 174, London.
83. Frederic Wood Jones to Arthur Keith, 24 August 1927. Correspondence between Sir Arthur Keith and Frederic Wood Jones, 1905–1951, no. 68, back page not numbered, The Royal College of Surgeons of England Archives, MSS Add 174, London.
84. Walter Baldwin Spencer to Frederic Wood Jones, 4 November 1925. Other Correspondence, The Royal College of Surgeons of England Archives, MSS 344 and 546.
85. From Wood Jones to Arthur Keith, 8 July 1923. Correspondence between Sir Arthur Keith and Frederic Wood Jones, 1905–1951, The Royal College of Surgeons of England Archives, MSS Add 174, London, no. 30, back page. For more criticism see Frederic Wood Jones to Arthur Keith, 27 February 1926. Correspondence between Sir Arthur Keith and Frederic Wood Jones, 1905–1951, no. 48, back page, The Royal College of Surgeons of England Archives, MSS Add 174, London.
86. Frederic Wood Jones, 'The Sandpiper', *Sea Birds Simplified*, Edward Arnold, London, 1934, p. 44.
87. Frederic Wood Jones to Arthur Keith, 10 December 1926. Correspondence between Sir Arthur Keith and Frederic Wood Jones, 1905–1951, The Royal College of Surgeons of England Archives, no. 62, MSS Add 174, London.
88. He wrote a colourful and self-serving autobiography; Stanley D. Porteus, *A Psychologist of Sorts: The Autobiography and Publications of the Porteus Maze Tests*, Pacific Books, Palo Alto, California, 1969. His daughter-in-law later wrote a hagiography; Elizabeth Dole, *Let's Go Exploring: the life of Stanley D. Porteus, Hawaii's pioneer psychologist*, Ku Pa'a, Honolulu, 1991.
89. B. E. Christophers, 'Frederic Wood Jones: his major books and how they were

reviewed', p. 650.

90. Frederic Wood Jones to Arthur Keith, 18 November 1928. Correspondence between Sir Arthur Keith and Frederic Wood Jones, 1905–1951, no. 75, back and front pages, The Royal College of Surgeons of England Archives, MSS Add 174, London.
91. Frederic Wood Jones to Arthur Keith, 3 January 1929. Correspondence between Sir Arthur Keith and Frederic Wood Jones, 1905–1951, no. 79, back page, The Royal College of Surgeons of England Archives, MSS Add 174, London.
92. Interview with James Guest, 3 June 2004.
93. Australasian Medical Congress, *Transactions of the Eleventh Session held in Brisbane, 1920*, Brisbane, Government Printer, 1921, p. 150. See Andrew Newton, The History of the Royal Australasian College of Surgeons from Foundation to 1935, unpublished BA (Hons) thesis, Monash University, 1972.
94. I discuss this in greater detail in chapter seven.
95. Frederic Wood Jones to Arthur Keith, 21 April 1930. Correspondence between Sir Arthur Keith and Frederic Wood Jones, 1905–1951, no. 94, back page to no. 95, The Royal College of Surgeons of England Archives, MSS Add 174, London.
96. Frederic Wood Jones to Arthur Keith, 15 April 1931. Correspondence between Sir Arthur Keith and Frederic Wood Jones, 1905–1951, no. 115, back page to 116, The Royal College of Surgeons of England Archives, MSS Add 174, London.
97. *Medical Journal of Australia*, 23 June 1917, p. 536.
98. Frederic Wood Jones, *Unscientific Excursions*, Butler and Tanner, Great Britain, 1934, p. 41.
99. ibid., pp. 43–4. For a broader discussion see Elazar Barkan, *The Retreat of Scientific Racism*.
100. Frederic Wood Jones, *Australia's Vanishing Race*, Angus & Robertson, Sydney, 1934, p. 22.
101. ibid., p. 21.
102. Jared Diamond, *Guns, Germs and Steel: the Fates of Human Societies*, Norton, New York, 1997. The germ of this idea can be found in the writings of the co-founder of the theory of natural selection, Alfred Russel Wallace. See A. R. Wallace, 'On the Varieties of Man in the Malay Archipelago,' 1865, *Transactions of the Ethnological Society of London*, 3, p. 206.
103. Frederic Wood Jones, *Australia's Vanishing Race*, pp. 16–7.
104. Frederic Wood Jones, *Life and Living*, p. 149. For the controversy see *Argus*, 10 December 1937.
105. James Guest, *Vicary Lecture, 1989*, 'John Hunter's disciple – Frederic Wood Jones', The Royal College of Surgeons of England, London, 1989, p. 7.
106. Frederic Wood Jones, 'Anatomy and a Life Principle', p. 121.
107. Frederic Wood Jones, *Habit and Heritage*, p. 100.

CHAPTER SEVEN (PAGES 163–197)

1. *Speculum*, no. 64, December 1905, p. 99.
2. ibid., no. 73, December 1908, p. 162.

3. ibid., no. 65, May 1906, p. 6.
4. ibid., no. 67, December 1906, pp. 120f.
5. K. F. Russell, *The Melbourne Medical School, 1862–1962*, Melbourne University Press, South Carlton, 1977, p. 105.
6. Anonymous, 'A Dissecting Room Wail', *Speculum*, no. 65, May 1906, pp. 22–3.
7. ibid., 'Grey and Cunny' refers to the two textbooks, Grey and Cunningham.
8. *Speculum*, no. 71, May 1908, p. 9.
9. K. F. Russell, op.cit., p. 105.
10. *Speculum*, no. 71, May 1908, p. 9. See also *Speculum*, no. 73, December 1908, pp. 148–9.
11. ibid., p. 9.
12. ibid.
13. The complaints continued as well in *Speculum*, no. 93, July, 1915, pp. 191–4:

    'This room here is reserved for the prosectors,
    Who bluffed Prof. Berry once in an exam.;
    Their job's like that of company directors—
    Well paid for doing little else but sham.'
14. K. F. Russell, op.cit., p. 105.
15. *Speculum*, no. 73, December 1908, p. 163.
16. ibid., no. 75, October 1909, p. 103.
17. ibid., no. 76, December 1909, p. 150.
18. ibid. The nasion can loosely be described as the nasal-bone and the bregma the front middle-point of skull, thus inferring a large forehead. Capilli of course refer to his impressive head of hair. Cunningham, Gray, Quain and Morris were anatomy texts used by the students. The facial index was a feature of the discipline of physical anthropology and was used to measure race and intelligence.
19. Berry lectured to dental and massage (later physiotherapy) students. A considerable gap in this book is the lack of treatment of these non-medical anatomy students. Word limitations necessitated their omission. They deserve a study of their own.
20. Wood Jones to Keith, 21 April 1930. Correspondence between Frederic Wood Jones and Sir Arthur Keith, 1905–1951, The Royal College of Surgeons of England Archives, MSS Add 174, London. In 1920, at the commencement of one of the earliest sociological and political studies of Australia, Professor Meredith Atkinson, Head of the Department of Sociology in the University of Melbourne, claimed that due to the excessive work loads demanded of university teaching in Australia '(p)robably no country in the world has been more forward in social experiment than Australia, and none so backward in contributing to the world stream of sociological thought.' Meredith Atkinson (ed), *Australia – Economic and Political Studies*, Melbourne, Macmillan, 1920, preface. For a discussion see Helen Bourke, 'Sociology and the Social Sciences in Australia, 1912–1928', *Australian and New Zealand Journal of Sociology*, vol. 17, 1981, p. 27; also Geoffrey Serle, *From Deserts the Prophets Come – The Creative Spirit in Australia 1788–1972*, Heinemann, Melbourne, 1973, p. 151; and Stephen Alomes, 'Intellectuals as

Publicists' in Brian Head and James Walter, *Intellectual Movements and Australian Society*, Oxford University Press, Melbourne, 1988, pp. 80–81.

21. R. J. A. Berry, 'The Modern University and its Upkeep' in Henry Payne (ed) *Proceedings of the First Educational Congress*, State of Victoria, held in Melbourne, 1912, Mullet, Melbourne, [no date], pp. 118–41. Berry later repeated the same message in R. J. A. Berry, 'The Present Needs and Future Requirements of the Medical School of the University of Melbourne', in *University of Melbourne, Medical School Jubilee* 1914, pp. 78–82.
22. The Federal Government did not at this time have the constitutional power, nor the finances, to fund education. This was to come during the Menzies' period in the 1950s and 60s.
23. R. J. A. Berry, 'The Modern University and its Upkeep', p. 122.
24. ibid., pp. 121; 122.
25. ibid., p. 123.
26. ibid.
27. ibid., p. 125.
28. ibid., p. 126.
29. ibid.
30. ibid.
31. *Age*, 18 March 1912; *Argus*, 18 March 1912.
32. Ronald Ridley (ed), *The Diary of a Vice-Chancellor*, p. 144.
33. ibid., 'Memorandum the University Council on the Financial Situation, 21 June, 1937', p. 348. I have not written extensively about the dental students in the Anatomy Department due to word limits.
34. ibid., 17 June 1937, p. 345.
35. K. F. Russell, op.cit., p. 109; R. J. A. Berry, *The Stewart Lectures*, *The Modern Psychology*, 14, 18 and 21 November, 1921, Robertson & Mullins, Melbourne, 1921.
36. K. F. Russell, op.cit., p. 114.
37. ibid., pp. 128–9.
38. ibid., pp. 134, 138, 139.
39. Sir Sydney Sunderland Papers, Acc 96/35, University of Melbourne Archives, University of Melbourne, file undated anonymous address, 'A. E. Coates – Master Surgeon', pp. 1, 9.
40. K. F. Russell, op.cit., p. 135.
41. ibid., p. 106.
42. R. J. A. Berry, 'Department of Anatomy and Histology, University of Melbourne Medical School', *Methods and Problems of Medical Education,* Seventeenth Series, The Rockefeller Foundation, New York, 1930, pp. 8–9.
43. ibid.
44. ibid.
45. ibid.

46. ibid.
47. K. F. Russell, op.cit., p. 87.
48. ibid., p. 110.
49. *Speculum*, no. 76, December 1909, p. 137.
50. ibid., p. 138.
51. ibid., p. 149.
52. *Speculum*, no. 83, May 1912, p .10; *Speculum*, no. 86, May 1913, p. 7.
53. ibid., no. 88, December 1913, p. 153.
54. ibid., no. 99, September 1917, p. 125; K. F. Russell, *The Melbourne Medical School*, p. 115.
55. *Speculum*, no. 139, May 1937, p. 13, editorial.
56. ibid., no. 107, November, 1920, pp. 85–6 in an interview with the Medical School porter, 'Fallopius', he remarked that '...'course Norman (Lindsay) done 'is anatomy 'ere same as all artists orter, and 'e made me a little present of these 'ere portraits.'
57. ibid., no. 104, October 1919, pp. 1–2.
58. T. S. Pensabane, *The Rise of the Medical Practitioner in Victoria*, pp. 72–5.
59. James A. Gillespie, *The Price of Health: Australian Governments and Medical Politics 1910–1960*, Cambridge University Press, Cambridge, 1991, chapter 1.
60. For example, in 1891 a private member's bill in the Legislative Assembly proposing a free university in the legislative assembly was defeated. See John Rickard, *H.B. Higgins: the Rebel as Judge*, George Allen & Unwin, Sydney, 1984, p. 155.
61. R. J. A. Berry, 'The Modern University and its Upkeep', op.cit., p. 129.
62. *Age*, 6 June 1917.
63. R. J. A. Berry, *Chance and Circumstance*, p. 102.
64. ibid., p. 103.
65. Ernest Scott, *A History of the University of Melbourne*, Melbourne University Press, Melbourne, 1936, pp. 214–17; Geoffrey Blainey, *A Centenary History of the University of Melbourne*, Melbourne University Press, Melbourne, 1957, pp. 142–5.
66. Ernest Scott, op.cit., p. 216.
67. ibid., pp. 213–14, 217.
68. 'The University of Melbourne: Annual Report', 1925–26, pp. 3–4, *Victorian Parliamentary Papers*, vol. II, 1926, pp. 1155–6. In 1927, the Education Department nominated 266, University High School 9, the University of Western Australia 1, the Railways Commissioners 3 and the War Bursars 8, making the number of 'free' places 287 in a student population of 2,985 or 9.6 per cent, 'The University of Melbourne: Annual Report, 1927–28, p. 1, *Victorian Parliamentary Papers*, vol. II, 1928, p. 1181.
69. See the *Melbourne University Calendar*, Melbourne University Press, for the years from 1930 to the present. The earlier figures can be found in the *Victorian Parliamentary Papers*.
70. R. J. A. Berry, *Chance and Circumstance*, p. 156.
71. ibid., p. 155.
72. ibid.

73. They then passed on to the womens' quarters, *Speculum*, no. 103, May 1919, p. 63.
74. R. J. A. Berry, *Chance and Circumstance*, p. 156, emphasis original.
75. ibid.
76. Sydney Sunderland, 'The Melbourne Medical School and some of its "characters" 1931–1975', *Chiron: Journal of the University of Melbourne Medical Society*, vol. 2, no. 2, 1992, p. 46.
77. R. J. A. Berry, op.cit., p. 157.
78. K. F. Russell, op.cit., pp. 129–30.
79. ibid., p. 130. For the plan see R. J. A. Berry, 'Department of Anatomy and Histology, University of Melbourne Medical School', op.cit.
80. William Henry Welch (1850–1934) was indicative of this approach. After conducting research in Germany he returned to the US to set up a pathology laboratory and, in 1900 he became the Professor of Pathology at Johns Hopkins University. He built up a fine research school as well as discovering the anaerobic bacteria responsible for gas gangrene.
81. R. J. A. Berry, *Through the States with a Seeing Eye*, Bristol, 1930.
82. B. T. Zwar to R. J. A. Berry, 8 January 1945 in R. J. A. Berry's archive *DDX 76*. Original documents held by the Lancashire Record Office, Preston, Lancashire, England.
83. James Barrett, 'the Halford Oration' in *Medical Journal of Australia*, December 10 1932. Berry's version in R. J. A. Berry, *Chance and Circumstance*, pp. 153–5.
84. *Speculum*, no. 71, May 1908, pp. 31–2.
85. *Farrago*, 17 April 1925, p. 1.
86. R. J. A. Berry, *Practical Anatomy*, three volumes, Robertson and Mullins, Melbourne, 1922, second ed. The date of the first edition seems to be unknown.
87. Sydney Sunderland, 'The Melbourne Medical School and some of its "characters" 1931–1975', p. 48.
88. *Speculum*, no. 125, November 1929.
89. R.J.A. Berry's archive *DDX 76*. See menus with place settings and signatures. R. J. W. Selleck, *The Shop*, p. 595.
90. Ronald Ridley (ed), *The Diary of a Vice-Chancellor*, Wednesday 8 December 1937, pp. 408–9.
91. David Clements Jackson, *The Six Horseshoes: Memoirs of a Personal and Professional Life*, Department of Child Health, University of Queensland, Brisbane, 1987, pp. 69–70. For the importance of dissecting in the first six decades of the twentieth century at least, interview with James Guest, 3 June 2004.
92. Interview with James Guest, 3 June 2004. See also interview with Gordon Keys Smith, 30 March 2006. The 'weekly treat' in *Speculum*, no. 140, April 1938, p. 55.
93. David Clements Jackson, *The Six Horseshoes*, pp. 69–70. See also interview with Gordon Keys Smith, 30 March 2006.
94. Wood Jones to Keith, 15 April 1931. Correspondence between Frederic Wood Jones and Sir Arthur Keith, 1905–1951, The Royal College of Surgeons of England Archives, MSS Add 174, London.

95. Wood Jones to Keith, 9 November 1936. Correspondence between Frederic Wood Jones and Sir Arthur Keith, 1905–1951, The Royal College of Surgeons of England Archives, MSS Add 174, London.
96. K. F. Russell, loc.cit., p. 153; David Clements Jackson, loc.cit., pp. 69–70.
97. Frederic Wood Jones to Keith, 15 April 1931. Correspondence between Frederic Wood Jones and Sir Arthur Keith, 1905–1951, The Royal College of Surgeons of England Archives, MSS Add 174, London.
98. J. Pearn, 'Hugo Flecker's contributions to medicine and natural history', Department of Child Health, Royal Children's Hospital, Brisbane, Qld.
99. See James Guest, *Vicary Lecture*, p. 6. The College was established in 1927 as the 'College of Surgeons of Australasia.' It was granted a Royal Charter and changed its name in March 1931.
100. Australasian Medical Congress, *Transactions of the Eleventh Session held in Brisbane, 1920*, Brisbane, Government Printer, 1921, p. 150. See Andrew Newton, The History of the Royal Australasian College of Surgeons from Foundation to 1935.
101. K. F. Russell, op.cit., p. 154.
102. Frederic Wood Jones to Keith 15 April 1931. Correspondence between Frederic Wood Jones and Sir Arthur Keith, 1905–1951, The Royal College of Surgeons of England Archives, MSS Add 174, London.
103. *Speculum*, no. 104, October 1919, p. 13.
104. Sydney Sunderland, 'The Melbourne Medical School and some of its "characters" 1931–1975', p. 98. See also, Gordon Keys Smith, Memoirs, unpublished manuscript in possession of author.
105. Wood Jones to Keith, 19 February 1931. Correspondence between Frederic Wood Jones and Sir Arthur Keith, 1905–1951, The Royal College of Surgeons of England Archives, MSS Add 174, London.
106. ibid.
107. Wood Jones to Keith, 15 April 1931. Correspondence between Frederic Wood Jones and Sir Arthur Keith, 1905–1951, The Royal College of Surgeons of England Archives, MSS Add 174, London.
108. R. J. W. Selleck, op.cit., pp. 658–9.
109. *Lady Julia Percy Island, 1935 expedition*, McCoy Society for Field Investigation and Research Melbourne, 1937, Melbourne, Government Printer; *Sir Joseph Banks Islands*: reports of the expedition of the McCoy Society for field investigation and research, Melbourne, Royal Society of Victoria, 1938. Interview with James Guest, 29 July 2004.
110. Wood Jones to Keith, 9 November 1936. Correspondence between Frederic Wood Jones and Sir Arthur Keith, 1905–1951, The Royal College of Surgeons of England Archives, MSS Add 174, London. See also Ronald Ridley (ed), *The Diary of a Vice-Chancellor, University of Melbourne, 1935–1938, Raymond Priestley*, Melbourne University Press, Carlton South, 2002, pp. 125, 149, 256, 270.
111. Ronald Ridley (ed), op.cit., p. 119; also p. 141.
112. Wood Jones to Keith, 10 December 1926. Correspondence between Frederic Wood Jones and Sir Arthur Keith, 1905–1951, The Royal College of Surgeons of England Archives, MSS Add 174, London.

113. Frederic Wood Jones, *Life and Living*, Kegan Paul, London, 1939, pp. 111–2.
114. ibid., p. 114.

CHAPTER EIGHT (PAGES **199–241**)

1. Letter of reference from Sir Alan Newton, 17 September 1936, Sir Sydney Sunderland papers, Acc 96/35, folder A0028, University of Melbourne Archives, The University of Melbourne.
2. Ian Darian-Smith, 'Sydney Sunderland 1910–1993', *Historical Records of Australian Science*, vol. 11, no. 1, 1996, p. 2.
3. Sir Sydney Sunderland papers, Acc 96/35, unidentified newspaper cutting, 19 July 1938, folder A0024.
4. Registrar to Sunderland, 19 March 1938, Sir Sydney Sunderland papers, Acc 96/35, folder A0036.
5. Sunderland to Registrar, 25 July 1938, ibid. Wood Jones wrote to the Vice-Chancellor supporting Sunderland's request for support. See Vice-Chancellor to Sunderland, 9 February 1939, 'I had a letter from Wood Jones in which he tells me something of the difficulties about which you recently consulted him. You have not written to me about them, but I thought that it might perhaps help you if I told you that I agree most emphatically with the advice that Wood Jones gave you ... without hesitation write to me if there is anything else that the University can do to help you on the financial side.' Sir Sydney Sunderland papers, ibid.
6. Registrar to Sunderland, op.cit.; see also Vice-Chancellor to Sunderland, 27 September 1938, forwarding a £500 advance free of interest from 1940 salary; also a further loan requiring payment of interest from the University accountant to Sunderland on 6 March 1940; Sunderland owes £593.10.0 from his trip, balance of payments show that they did charge interest. Sir Sydney Sunderland papers, op.cit.
7. 'I am backing young Sunderland for the Chair in Hong Kong', Wood Jones to Keith, 28 July 1936. Correspondence between Frederic Wood Jones and Sir Arthur Keith, 1905–1951, The Royal College of Surgeons of England Archives, MSS Add 174, London.
8. ibid.
9. For their courtship *Table Talk,* 28 July 1938. For the relationship with Le Gros Clark see Ian Darian-Smith, 'Sydney Sunderland 1910–1993', *Historical Records of Australian Science*, vol. 11, no. 1, 1996, p. 2.
10. Sunderland to Registrar, op.cit.
11. Wood Jones to Keith, 8 November 1950; 'Dear Sir Arthur, I am sending Sunderland's R.[oyal] S.[ociety] Certificate, which you said you would be good enough to sign as his seconder. Walshe, Kellaway and Dale have so far promised and I hope to get the thing completed in time. Would you send it on, after signature, to Walshe with the enclosed note in the envelope sent with it. Ever yours with best wishes, F.W.J.' Correspondence between Frederic Wood Jones and Sir Arthur Keith, 1905–1951, op.cit. I have been told Le Gros Clark blocked it. Sunderland later arranged for Le Gros Clark to receive an honorary doctorate at Melbourne University in 1952. See

Le Gros Clark to Sunderland, 7 September 1952, Sir Sydney Sunderland papers, Acc 96/35, folder A0038.

12. Ian Darian-Smith, 'Sydney Sunderland 1910–1993', p. 3.
13. For a detailed discussion of the important experiences of Sunderland on this tour see ibid., p. 2.
14. Flexner to Sunderland, 21 January 1954, Sir Sydney Sunderland papers, Acc 96/35, folder A0038. Louis Barkhouse Flexner, founder of the Mahoney Institute of Neurological Research, University of Pennsylvania, was a world leader in the study of memory and learning. Part of a great US medical dynasty, his two uncles were Drs Simon Flexner and Abraham Flexner. The latter, as the author of a famous study on reform of medical education, The Flexner Report, was recognised as the father of modern medicine in the USA. The former became the first director of the Rockefeller Institute of Medical Research. For many other such letters see, for example, Sir Sydney Sunderland papers, op.cit., folders A0003 and A0038.
15. Sir Sydney Sunderland papers, address on received honorary doctorate at the first graduation ceremony of the Tasmanian Medical Faculty, December 1970, p. 3, folder A0030.
16. Petition from students, 2 April 1946, Sir Sydney Sunderland papers, folder A0006.
17. Discussions with Brian Egan and Keith Henderson, 1 and 2 February 2007.
18. Letter of recommendation by Sir Hugh B. Devine for Sunderland's application for Professorship at the University of Hong Kong. Sir Sydney Sunderland papers, folder A0028. For a summary of his work see *Speculum*, no. 140, April 1938, pp. 10–11.
19. Sunderland to Paton, 21 June 1952, Sir Sydney Sunderland papers, folder A0036.
20. Paton to Sunderland, 16 June 1952, Sir Sydney Sunderland papers, ibid.
21. Sunderland to Paton, op.cit.
22. ibid.
23. ibid.
24. ibid.
25. Medley to Sunderland, 1 July 1941, Sir Sydney Sunderland papers, ibid. See G. Serle, *Sir John Medley*, Melbourne University Press, Carlton South, 1993.
26. Paton to Sunderland, 6 November 1953, Sir Sydney Sunderland papers, folder A 0036.
27. Paton to Sunderland, 12 April1965, Sir Sydney Sunderland papers, folder A 0036.
28. John Poynter and Carolyn Rasmussen, *A Place Apart: The University: Decades of Challenge*, Melbourne University Press, Carlton South, 1996, p. 149.
29. Sunderland to Paton, 21 June 1952, Sir Sydney Sunderland papers, op.cit.
30. K. F. Russell, *The Melbourne Medical School: 1862–1962*, Melbourne University Press, Carlton South, 1977, p. 181.
31. Gordon Keys Smith, Extracts of Memoirs of Dr Gordon Keys Smith, who was a medical student under Professor Wood Jones, 1935–1936, and on the staff of the Anatomy Department under Professor Sunderland, 1946–1948. The Memoirs 'Reminiscences and Reflections', an unpublished manuscript, vol. 1, 2005; vol 2 in preparation, p. 5.

32. Erwin Schindler, 'Arthur Schüller: Pioneer of Neuroradiology', *American Journal of Neuro Radiology*, 18: August 1997, p. 1297.
33. J. W. D. Bull, 'History of neuroradiology', *British Journal of Radiology*, 34: 1961, pp. 69–84.
34. Erwin Schindler, 'Arthur Schüller: Pioneer of Neuroradiology', loc.cit. Gordon Keys Smith, Extracts of Memoirs, pp. 5–6. Discussions with Brian Egan and Keith Henderson, 1 and 2 February 2007.
35. Gordon Keys Smith, op.cit., p. 5.
36. ibid.
37. ibid.
38. Interview with James Guest, 29 July 2004.
39. See Obituary 'William Ross Adey, Professor of Anatomy and Physiology, Los Angeles, 1922–2004', http://www.universityofcalifornia.edu/senate/inmemoriam/williamrossadey.htm.
40. Geoffrey Kenny, 'John Ochiltree Lavarack: an obituary', *Chiron*, vol., 4, no. 2, May 1999, p. 49.
41. For his interest see boxes 14 and 15, Sir Sydney Sunderland papers, op.cit.
42. *Speculum*, no. 131, July, 1932, p. 90.
43. Sir Sydney Sunderland papers, op.cit., folders A0158, A0122.
44. *ISI Neuroscience Citation Index*, Ian Darian-Smith, op.cit., p. 5.
45. ibid., p. 4.
46. ibid.
47. Undated, untitled lecture on the wartime study of 300 peripheral nerve injuries, Sir Sydney Sunderland papers, op.cit., folder A0035.
48. ibid., pp. 1–2.
49. Ian Darian-Smith, op.cit., p. 5.
50. Undated, untitled lecture on the wartime study of 300 peripheral nerve injuries, op.cit., p. 5.
51. Ian Darian-Smith, loc.cit. For a more detailed description of his research see ibid., pp. 5–8.
52. Gordon Keys Smith, extracts of Memoirs, op.cit., p. 8.
53. Wood Jones to Keith, 22 November 1934. Correspondence between Frederic Wood Jones and Sir Arthur Keith, op.cit. Wood Jones was furious.
54. Harris to Sunderland, 9 October 1945, Sir Sydney Sunderland papers, folder A0038.
55. Harris to Sunderland, 17 January or February 1947; 7 March 1947. The Sunderlands sent the Harris family food parcels.
56. Keith to Sunderland, 4 December 1952, Sir Sydney Sunderland papers, folder A0038.
57. Keith to Sunderland, 3 January 1949; 9 December 1951; 22 May 1953; 20 January 1954; and more, Sir Sydney Sunderland papers, folder A0038.
58. Richard Evans, *The Pyjama Girl Mystery*, Scribe, Carlton North, 2004.
59. Interview with Ian Sunderland, 31 January 2007; interview with Keith Henderson, 2 February 2007.

60. Report for the Wellcome Trust for a grant of £250 to visit laboratories in the United Kingdom in June 1961, pp. 2–3. Sir Sydney Sunderland papers, folder A0005.
61. ibid., p. 3.
62. ibid., p. 4.
63. Ian Darian-Smith, op.cit, p. 10; Frank Fenner (ed), *The First Twenty-Five Years: The Australian Academy of Science*, the Academy, Canberra, 1980; Frank Fenner (ed), *The First Forty Years*: The Australian Academy of Science, Canberra, 1995; Sir Sydney Sunderland papers, folders A0110, A0111, A0112, A0138, A0146.
64. Ian Darian-Smith, op.cit., p. 11.
65. Penington to Sunderland, 19 March 1982, Sir Sydney Sunderland papers, folder A0038.
66. Morison to Sunderland, 11 August 1955, Sir Sydney Sunderland papers, folder A0038. Ross Adey is mentioned above. Sir John C. Eccles was a demonstrator in Berry's department and later, Nobel Laureate.
67. Fenner to Sunderland, 23 May 1956, Sir Sydney Sunderland papers, folder A0038.
68. Department of External Affairs to Vice-Chancellor of the Australian National University, 11 May 1956, Sir Sydney Sunderland papers, ibid.
69. Medical Personnel Needs of Community in Victoria, Summary of British Medical Association Investigation, 1955, Sir Sydney Sunderland papers, folder A0131.
70. Sunderland to Fenner, 31 May 1956, Sir Sydney Sunderland papers, folder A0038.
71. Fenner to Sunderland, 22 June 1956, Sir Sydney Sunderland papers, folder A0073.
72. Canadian Technical Education Report in Fenner to Sunderland, ibid.
73. ibid.
74. ibid.
75. Frank Fenner, Sydney Sunderland, H. N. Robson, Report on Medical Education in Indonesia, p. 10, Sir Sydney Sunderland papers, folder A0104.
76. ibid.
77. Bulletin of the Australian-Asian Association of Victoria, 1958, p. 20, Sir Sydney Sunderland papers, folder A0177.
78. Frank Fenner, *Nature, Nurture and Chance: The Lives of Frank and Charles Fenner*, 'Other Activities, 1953 to 1957', http://epress.anu.edu.au/nature/mobile_devices/index.html. They failed to visit the school in Medan.
79. Frank Fenner, Sydney Sunderland, H. N. Robson, op.cit., p. 6, Sir Sydney Sunderland papers, folders A0058 and A0104.
80. ibid., pp. 5–6.
81. ibid., p. 6.
82. ibid., p. 8.
83. ibid., p. 9.
84. ibid.
85. ibid., p. 11. For staff and facilities see pp. 10–12.
86. ibid.
87. ibid., p. 16.

88. ibid.
89. ibid., p. 18.
90. Sunderland to Bror Rexed, 20 July 1954, Sir Sydney Sunderland papers, folder A0109.
91. Frank Fenner, Sydney Sunderland, H. N. Robson, op.cit., p. 19, Sir Sydney Sunderland papers, folder A0104.
92. Sydney Sunderland, A Reply to Professor Wright's Statement on the Proposed Bukittingi Medical School Project, p. 1, 16 December 1957, Sir Sydney Sunderland papers, folder A0151.
93. ibid.
94. Letter from Professor Wright to the Vice-Chancellor with Statement, 12 December 1957, Sir Sydney Sunderland papers, folder A0151.
95. ibid.
96. As a motion of confidence in his performance the Faculty voted in Sunderland for five years in 1955. See John Poynter and Carolyn Rasmussen, op.cit., p. 156.
97. Letter from Professor Wright to the Vice-Chancellor with Statement, op.cit.
98. Sunderland to Council, 16 December 1957, Sir Sydney Sunderland papers, ibid.
99. ibid.
100. Proposed agreement between the University of Melbourne and the Commonwealth of Australia, 3 February 1958, Sir Sydney Sunderland papers, ibid.
101. As they continued to do when the Kuala Lumpur project eventuated. See voluminous correspondence, Sir Sydney Sunderland papers, folder A0179.
102. *Sun*, 12 December 1957.
103. Short biographical note, Sir Sydney Sunderland papers, folder A0005.
104. Australian Embassy to Department of External Affairs, 3 October 1958, Sir Sydney Sunderland papers, folder A0104.
105. ibid.
106. ibid.
107. Colin T. Moodie, Department of External Affairs to Sunderland, 23 April 1959, Sir Sydney Sunderland papers, ibid.
108. The story of the Melbourne Medical School in Kuala Lumpur is far too long to tell here, but deserves a detailed history of its own.
109. Calwell to Sunderland, 23 August 1956, Sir Sydney Sunderland papers, folder A0028.
110. Sunderland to Calwell, 7 September 1956, ibid.
111. Programme for the Wallaby Club 74th annual dinner, 5 July, 1968, Sir Sydney Sunderland papers, folder A0024.
112. James Le Fanu, *The Rise and Fall of Modern Medicine*, Abacus, London, 1999, p. xv.
113. For a full discussion of the introduction of quotas and the backlash see John Poynter and Carolyn Rasmussen, op.cit., pp. 154–64.
114. Medical Personnel Needs of Community in Victoria, Summary of British Medical Association Investigation, 1955, Sir Sydney Sunderland papers, folder A0131. For an

overview of the medical politics of the time see James Gillespie, *The Price of Health: Australian Governments and Medical Politics 1910–1960,* Cambridge University Press, 2002, pp. 280ff.

115. Medical Personnel Needs of Community in Victoria, ibid.
116. ibid.
117. University of Melbourne, Second Medical School, Statement by Sir Peter MacCallum, 5 May 1956, p. 3, Sir Sydney Sunderland papers, folder A0131.
118. ibid.
119. ibid.
120. Sydney Sunderland, Report on Overseas Experiences Relating to the Entry of Students into Medical Schools, 27 April 1954, p. 2, Sir Sydney Sunderland papers, folder A0073.
121. Sydney Sunderland, Policy relating to the admission of students to the medical course in the University of Melbourne, 21 July 1954, p. 2, Sir Sydney Sunderland papers, folder A0073. The figure of two hundred was reached by an analysis of pass rates. This was to create significant overcrowding, as recognised by the Medical Quota Committee, set up by the Council. Medical Quotas Committee, Memorandum to Council plus Appendix detailing overcrowding in the medical faculty. See also S. Sunderland, University of Melbourne, Melbourne Medical School, 22 November, 1962, p. 2; Sir Sydney Sunderland papers, folder A0073.
122. Sydney Sunderland, The Case for a Second Medical School in Victoria, 27 February 1956, Sir Sydney Sunderland papers, folder A0073.
123. Sydney Sunderland, University of Melbourne, Melbourne Medical School, 22 November, 1962, p. 1; Untitled document commencing 'In 1962 the University of Melbourne Medical School', Sir Sydney Sunderland papers, folder A0073.
124. Sydney Sunderland, Report on overseas experience relating to the entry of students into medical schools, op.cit., pp. 6–7. For similar criticisms see John Poynter and Carolyn Rasmussen, op.cit., p. 150.
125. Sydney Sunderland, University of Melbourne, Melbourne Medical School, 22 November 1962, pp. 1–2, Sir Sydney Sunderland papers, folder A0073.
126. John Poynter and Carolyn Rasmussen, op.cit., p. 156.
127. *The Age,* 15 October 1959. See John Poynter and Carolyn Rasmussen, ibid., pp. 154–6.
128. *Age*, 15 October 1959.
129. ibid.
130. Sydney Sunderland, University of Melbourne, loc.cit., Sir Sydney Sunderland papers, folder A0073.
131. ibid.
132. Sunderland to Paton, 6 February 1961, S. Sunderland, ibid., p. 3, Sir Sydney Sunderland papers, folder A0036.
133. Sydney Sunderland, Chair of Experimental Neurology: Conditions of Acceptance of New Appointment and Resignation of Professorship of Anatomy, undated memorandum, Sir Sydney Sunderland papers, folder A0036.

134. ibid.
135. Sunderland to Ray, 15 February 1971, Sir Sydney Sunderland papers, folder A0076.
136. Paton to Sunderland, 12 July 1962, Sir Sydney Sunderland papers, folder A0036.
137. R. D. Wright, Dean's lecture, *Chiron*, March 1984, p. 9.
138. ibid.
139. Menzies to Sunderland, 2 April 1962; Sunderland to Menzies, 3 April 1962, Sir Sydney Sunderland papers, folder A0038; *Sun*, 21 June 1962.
140. Latrobe University was established by an Act of Parliament in 1964 and admitted its first students in 1967.
141. Menzies to Martin, 31 March 1966, Sir Sydney Sunderland papers, folder A0038.
142. Sunderland to Bolte, 8 August 1963, Sir Sydney Sunderland papers, folder A0094.
143. ibid. Copy of the faculty statement of 28 August 1963, Sir Sydney Sunderland papers, folder A0074.
144. John Poynter and Carolyn Rasmussen, op.cit., p. 151; for an extended discussion of this expansion see pp. 249–54.
145. Minutes of Melbourne University Council Meeting no. 15, 2 October 1963, note 25, Sir Sydney Sunderland papers, folder A0074.
146. John Poynter and Carolyn Rasmussen, op.cit., p. 247.
147. ibid., p. 249.
148. *Farrago*, vol. 40, no. 14, 6 July 1964, p. 1. John Poynter and Carolyn Rasmussen, ibid., pp. 252–3.
149. Handwritten notes for unnamed meeting, Sir Sydney Sunderland papers, folder A0074.
150. *Farrago*, vol. 40, no. 14, 6 July 1964, p. 1.
151. ibid.
152. ibid.
153. ibid.
154. John Poynter and Carolyn Rasmussen, op.cit., pp. 263f.
155. A number of reports of committees, including the Professorial Board detailing the woes, Sir Sydney Sunderland papers, folder A0074.
156. Inspection of Medical Centre and buffet, 13 June 1968, Sir Sydney Sunderland papers, folder 0024.
157. It is not possible in this work to list all the organisations Sunderland belonged to. For a full list see Ian Darian-Smith, op.cit.
158. Sunderland to Bror Rexed, 20 July 1954, Sir Sydney Sunderland papers, folder A0109.

CHAPTER NINE (PAGES 243–251)

1. John Poynter and Carolyn Rasmussen, *A Place Apart: The University: Decades of Challenge*, Melbourne University Press, Carlton South, 1996, p. 281. This period of the history of the Medical School, beginning with the move into the new building in 1968, has not yet run its course. Many involved are still active and any evaluation

would be incomplete. This chapter, therefore, is much less of a history of the Department. Rather, most of this chapter is the author's summary of the accounts of a number of important figures in the Department in these years. Those who provided information are Graeme Ryan, John Furness, Daine Alcorn, Antony Goodwin, Bernice Stratford, Ian Darian-Smith, Tony Buzzard and Geoff Kenny. Many more would have been consulted if the full story of the period were to be told. The transcripts of their interviews can be accessed in the Anatomy and Cell Biology Department Archives. This chapter would also not have been possible without the considerable assistance of Chris Briggs.

2. John Furness, Transforming the Department: 1978–2006, unpublished manuscript, 2006, Anatomy and Cell Biology Department Archives.
3. ibid.
4. Interview with Bernice Stratford, 6 February 2007.
5. Gordon Keys Smith, extracts of Memoirs of Dr Gordon Keys Smith, who was a medical student under Professor Wood Jones, 1935–36, and on the staff of the Anatomy Department under Professor Sunderland, 1946–48. The Memoirs 'Reminiscences and Reflections' is an unpublished manuscript, vol. 1, 2005; vol. 2 in preparation, pp. 8–9.
6. George Halford, G. H. Fetherston, Patrick Moloney, J. E. Neild, William Rees, Alexander Mackie.
7. Frederic Wood Jones, *Life and Living*, Kegan Paul, London, 1939, p. 114.

Heads of 'Department' of Anatomy

George Britton Halford 1862–1881

Harry Brookes Allen 1882–1906

Richard J A Berry 1906–1929

Frederic Wood Jones 1930–1937

Eustace Alwynne Rowlands 1938–1939*

Sydney Sunderland 1940–1961

Leslie Ray 1961–1977

Graeme Ryan 1978–1984

Ian Darian-Smith 1985–1992

John Barton Furness 1993–96; 1999

Daine Alcorn 1996–98; 2000–2002

Antony Goodwin 2002–

*Acting Head – others who acted for shorter periods have not been included in this list

Select bibliography

Archival sources

Alma Mater: A University Newspaper.

'Brownless Memorial Supplement, Being a Special Medical Supplement' in *Alma Mater*, vol. iii, no. 6, September 1898.

Arthur Keith Papers, The Royal College of Surgeons of England Archives, London.

Australian Medical Journal.

Berry, R. J. A., *Chance and Circumstance*, unpublished autobiography, University of Melbourne Medical Library, 1954.

Cadaver Archives, Department of Anatomy and Cell Biology, University of Melbourne, Parkville.

Furness, John, Transforming the Department: 1978–2006, unpublished manuscript, Department of Anatomy and Cell Biology Archives, 2006.

Harry Brookes Allen Papers, University of Melbourne Archives, University of Melbourne.

R. J. A. Berry Archive DDX 76I, Lancashire Record Office, Preston, Lancashire, England.

Sir Sydney Sunderland Papers, University of Melbourne Archives, University of Melbourne.

Speculum: a Journal of the Melbourne Medical Students.

Wood Jones Papers, The Royal College of Surgeons of England Archives, London

Texts, theses and journal articles

Allen, H. B., 'A History of the Medical School' in *University of Melbourne, Medical School Jubilee 1914*, Carlton.

Alomes, Stephen, 'Intellectuals as Publicists' in Brian Head and James Walter, *Intellectual Movements and Australian Society*, Oxford University Press, Melbourne, 1988.

Anderson, Warwick, *The Cultivation of Whiteness: Science, Health and Racial Destiny in Australia,* Melbourne University Press, Carlton South, 2002.

Atkinson, Meredith (ed), *Australia – Economic and Political Studies*, Melbourne, Macmillan, 1920.

Barkan, Elazar, *The Retreat of Scientific Racism: Changing Concepts of Race in Britain and the United States between the Two World Wars*, Cambridge University Press, Cambridge, 1992.

Barrett, James, 'The Halford Oration', *Medical Journal of Australia*, 10 December 1932, p. 708.

Beaney, J. G., *Doctors Differ: A Lecture*, Melbourne, 1876.

Beard, Chris, *The Hunt for the Dawn Monkey: Unearthing the Origins of Monkeys, Apes and Humans*, University of California Press, Berkeley, 2004.

Berry, R. J. A. and Büchner, L. W. G., 'The Correlation of Size of Head and Intelligence as Estimated from the Cubic Capacity of the Brain of 355 Melbourne Criminals' [read 10 October, 1912], *Proceedings of the Royal Society of Victoria*, 25 (N.S.), Pt. II., 1913, p. 229.

Berry, R. J. A., and Gordon R. G., *The Mental Defective: A Problem in Social Inefficiency*, Kegan Paul, London, 1931.

Berry, R. J. A., and Porteus, S. D., 'Size of Brain and Intelligence', *Medical Journal of Australia*, 18 August 1917, p. 155.

Berry, R. J. A., and Robertson, A. W. D., *The place in nature of the Tasmanian Aboriginal as deduced from a study of his calvaria, Part 2, His relation to the Australian Aboriginal*, Edinburgh, Robert Grant & Son, 1914.

Berry, R. J. A., *Report to the Edward Wilson (of 'The Argus') Trust on Mental Deficiency in the State of Victoria, with Suggestions for the Establishment of a Child Guidance Clinic*, Argus, Wilson and Mackinnon, Melbourne, 1929.

———., 'Department of Anatomy and Histology, University of Melbourne Medical School', *Methods and Problems of Medical Education*, Seventeenth Series, The Rockefeller Foundation, New York, 1930, p. 8.

———., *Practical Anatomy*, three volumes, Robertson and Mullins, Melbourne, 1922, second ed. The date of the first edition seems to be unknown.

———., *The Stewart Lectures, The Modern Psychology*, 14, 18 and 21 November, 1921, Robertson & Mullins, Melbourne, 1921.

———., *Through the States with a Seeing Eye*, Bristol, 1930.

Blackburn, Kevin, 'Preaching "The Gospel of Efficiency": The Promotion of Ideas about Profit-sharing and Payment by Results in Australia, 1915–1929', *Australian Historical Studies*, vol. 27, no. 107, October 1996, p. 257.

Blainey, Geoffrey, *A Centenary History of the University of Melbourne*, Melbourne University Press, Carlton South, 1957.

Bourke, Helen, 'Sociology and the Social Sciences in Australia, 1912–1928', *Australian and New Zealand Journal of Sociology*, vol. 17, 1981, p. 27.

Bowler, Peter J., *Evolution: The History of an Idea*, University of California Press, Berkeley, 1989.

Brennan, E. J. T. (ed), *Education for National Efficiency: the Contribution of Sidney and Beatrice Webb*, Athlone Press, London, 1975.

Brown, Ralph, *A Delineation of the Character, Talents, Physiological Developments and Natural Adaptations of M.*, Melbourne, 1894.

Butcher, Barry W., 'Darwinism, Social Darwinism, and the Australian Aborigines: A Reevaluation' in MacLeod, Roy and Rehbock, Philip (eds), *Evolutionary Theory and Natural History in the Pacific: Darwin's Laboratory*, University of Hawaii Press, Honolulu, 1994.

———. Darwinism and Australia: 1836–1914, unpublished PhD thesis, University of Melbourne, 1992.

———. 'Gorilla Warfare in Melbourne: Halford, Huxley and 'Man's place in nature',

Home, R.W. (ed), *Australian Science in the Making*, Cambridge University Press, Cambridge, 1988.

Cage, R. A., *Poverty Abounding, Charity Aplenty: the charity network in colonial Victoria*, Hale & Iremonger, Sydney, 1992.

Carter, J. M. T., *Painting the Islands Vermilion: Archibald Watson and the brig Carl*, Melbourne University Press, Carlton South, 1999.

Cawte, Mary, 'Craniometry and Eugenics in Australia: R. J. A. Berry and the Quest for Social Efficiency', *Historical Studies*, vol. 22, no. 86, April 1986, p. 35.

Christophers, B. E., 'Frederic Wood Jones: Coral and Atolls', *Australian and New Zealand Journal of Surgery*, no. 65, 1995, p. 752.

———. 'Frederic Wood Jones: His Illnesses', *Australian and New Zealand Journal of Surgery*, no. 70, 2000, p. 526.

———. 'Frederic Wood Jones: His major books and how they were reviewed', *Australian and New Zealand Journal of Surgery*, no. 67, 1997, p. 646.

Clark, Wilfrid E. Le Gros, 'Wood Jones, Frederic 1879-1954', *Biographical Memoirs of Fellows of the Royal Society*, vol. 1, November 1955.

Craig, C., 'The Egregious Dr Beaney of the Beaney Scholarships', *Medical Journal of Australia*, 6 May 1950, p. 593.

Crook, Paul, *Darwinism, War and History: The debate over the biology of war from the 'Origin of Species' to the First World War*, Cambridge University Press, Cambridge, 1994.

Cumpston, J. H. L., *The Health of the People: A study in federalism*, Roebuck, Canberra, 1978.

Dalrymple, William, *White Mughals: Love and Betrayal in Eighteenth Century India*, Viking, London, 2003.

Darian-Smith, Ian, 'Sydney Sunderland 1910–1993, *Historical Records of Australian Science.*

Dawson, Warren R. (ed), *Sir Grafton Elliot Smith: a biographical record: by his colleagues*, Cape, London, 1938.

Desmond, Adrian and Moore, James, *Darwin*, Penguin, London, 1991.

Desmond, Adrian, *Huxley, The Devil's Disciple*, Michael Joseph, London, 1994.

Diamond, Jared, *Guns, Germs and Steel: the fates of human societies*, Norton, 1997, New York.

Dixon, Joan M., 'Melbourne 1865: Gorillas at the Museum', in Carolyn Rasmussen, *A Museum for the People: A History of Museum Victoria and its predecessor institutions, 1854–2000*, Carlton North, Scribe Publications, 2001.

Dole, Elizabeth, *Let's Go Exploring: the life of Stanley D. Porteus, Hawaii's pioneer psychologist*, Ku Pa'a, Honolulu, 1991.

Egan, Bryan, *Ways of a Hospital: St Vincent's Hospital 1890s to 1990s*, Allen & Unwin, St. Leonards, 1993.

Elkin, A. P., (ed), *Grafton Elliot Smith: the Man and his Work*, Sydney University Press, Sydney, 1974.

Evans, Richard, 'Epidemics and Revolutions', in Terence Ranger and Paul Slack (eds), *Epidemics and Ideas,* Cambridge University Press, Cambridge, 1993.

———. *The Pyjama Girl Mystery*, Scribe, Carlton North, 2004.

French, Roger, 'The Anatomical Tradition' in W. F. Bynum and Roy Porter, *Companion Encyclopedia of the History of Medicine*, Routledge, London, 1997, vol. 1.

Gandevia, Bryan, *The Melbourne Medical Students 1862–1942*, Melbourne Medical Students' Society, Melbourne, 1948.

———. 'James Thomas Rudall, FRCS (1828–1907): His Life, and Journal for the Year 1858', *Medical Journal of Australia*, 25 December 1954, p. 992.

———. 'Some Aspects of the Life and Work of James George Beaney', *Medical Journal of Australia*, 2 May 1953, p. 614.

Garton, Stephen, 'Sound Minds and Healthy Bodies: Re-considering Eugenics in Australia, 1914–1940', *Australian Historical Studies*, vol. 26, no. 103, October 1994, p. 164.

Geary, Laurence, 'Australian Medical Students in 19th Century Scotland', *Proceedings of the Royal College of Physicians, Edinburgh*, vol. 26, 1996; 'The Scottish-Australian connection, 1850–1900', in Vivian Nutton and Roy Porter (eds), *the History of Medical Education in Britain*, Rodopi, Amsterdam, 1995.

Gillespie, James A., *The Price of Health: Australian Governments and Medical Politics 1910–1960*, Cambridge University Press, Cambridge, 1991, 2002.

Goddard, Henry Herbert, *The Kallikak Family: A Study in the Heredity of Feeble-mindedness*, Macmillan, New York, 1921.

Gould, Stephen Jay, *Bully for Brontosaurus: Reflections in Natural History*, Vintage, London, 2001.

Gregory, Alan, *The Ever Open Door: A History of the Royal Melbourne Hospital*, Hyland House, South Melbourne, 1998.

Guest, James, *Vicary Lecture, 1989*, 'John Hunter's disciple – Frederic Wood Jones', The Royal College of Surgeons of England, London, 1989.

Halford, G. B., 'On the skeleton of the gorilla', *Transactions of the Royal Society of Victoria*, vol. 7, 1866.

———. 'Quadrumanous or bimanous?', *Australian Medical Journal*, July 1863.

———. *Not Like Man, Bimanous and Biped; Nor Yet Quadrumanous; but Cheiropodus*, Wilson & Mackinnon, Melbourne, 1863.

Hardy, Anne, *The Epidemic Streets: infectious disease and the rise of preventive medicine,* Oxford University Press, Clarendon Press, New York, 1993.

Hearn, W. E., *Plutology, or the Theory of Efforts to Satisfy Human Wants*, G. Robertson, Melbourne, 1863.

Jablonka, Eva and Lamb, Marion J., *Epigenetic Inheritance and Evolution: The Lamarckian Dimension*, Oxford University Press, Oxford, 1995.

Jackson, David Clements, *The Six Horseshoes: Memoirs of a Personal and Professional Life*, Department of Child Health, University of Queensland, St. Lucia, 1987.

Jensen, J. Vernon, 'Return to the Wilberforce-Huxley debate', *British Journal for the History of Science*, vol. 21, 1988.

Jones, Ross L., 'The Master Potter and the Rejected Pots: Eugenic Legislation in Victoria 1918–1939', *Australian Historical Studies*, vol. 30, no. 113, October 1999.

———. 'The Rogers-Templeton and Pearson Royal Commissions: Contemporary Views of the 1872 Education Act', *History of Education Review*, vol. 27, no. 2, 1998.

Jones, William Ernest, *Report on Mental Deficiency in the Commonwealth of Australia*, H.J. Green, Government Printer, Canberra, 1929.

Kehoe, Mary, *The Melbourne Benevolent Asylum: Hotham's Premier Building: the Annals of Hotham, volume 1*, Hotham History Project, North Melbourne, 1998.

Keith, Arthur, *An Autobiography*, Watts, London, 1950.

Kenny, A. L., 'Halford the Man: The Halford Oration', *Medical Journal of Australia*, 1 February 1941, p. 130.

Kenny, Geoffrey, 'John Ochiltree Lavarack: an obituary', *Chiron*, vol. 4, no. 2, May 1999, p. 49.

Kevles, Daniel, *In the Name of Eugenics*, Harvard University Press, London, England, 2001 (1985).

Kiernan, V. G., *The Lords of Human Kind: black man, yellow man, and white man in an age of empire*, London, Cresset Library, 1988.

Kisch, Egon Erwin, *Australian Landfall*, translated by John Fisher & Irene and Kevin Fitzgerald, Australasian Book Society, Sydney, 1969 (1937).

Lady Julia Percy Island, 1935 expedition, McCoy Society for Field Investigation and Research Melbourne, 1937, Melbourne, Government Printer; *Sir Joseph Banks Islands*: reports of the expedition of the McCoy Society for field investigation and research, Melbourne, Royal Society of Victoria, 1938.

Lewis, John, 'So Much Grit in the Hub of the Educational Machine: Schools, Society and the Invention of Measurable Intelligence' in Bob Bessant (ed), *Mother State and Her Little Ones: Children and Youth in Australia 1860s to 1930s*, Centre for Youth and Community Studies, Melbourne, 1987.

Lewis, John, Removing the Grit: The Development of Special Education in Victoria 1887–1947, unpublished PhD thesis, LaTrobe University, 1989.

Love, Harold, *James Edward Neild, Victorian Virtuoso*, Melbourne University Press, Carlton 1989.

Lovejoy, Arthur O., *The Great Chain of Being: A Study of the History of an Idea*, Harper, New York, 1960.

MacDonald, H. P., *Human Remains: episodes in human dissection*, Melbourne University Press, Carlton South, 2005.

Macmillan, Malcolm, 'Localisation and William Macewen's Early Brain Surgery Part I: The Controversy', *Journal of the History of the Neurosciences*, vol. 13, no. 4, December, 2004, p. 24.

Malik, Kenan, *The Meaning of Race: Race, History and Culture in Western Society*, New York University Press, New York, 1996.

McCalman, Janet, *Sex and Suffering: Women's Health and a Women's Hospital: The Royal Women's Hospital, Melbourne 1856–1996*, Melbourne University Press, Carlton South, 1998.

McPhee, Peter, *Pansy: a life of Roy Douglas Wright*, Melbourne University Press, Carlton, 1999.

Montagu, Francis M. Ashley, 'The Tarsian Hypothesis and the Descent of Man', *The Journal of the Royal Anthropological Society of Great Britain and Ireland*, vol. 60 (July to December, 1930), p. 335.

Morison, Patricia, *J. T. Wilson and the Fraternity of Duckmaloi*, Atlanta, Amsterdam, 1997.

Mozley, Ann 'Evolution and the Climate of Opinion in Australia, 1840–76' *Victorian Studies*, vol. 10, June 1967.

Neild, James E., 'Presidential Address', *Australasian Medical Gazette*, January 1901, p. 6.

———. 'The Medical School of the Melbourne University: part II', *Australian Medical Journal*, vol. ix, no. 6, 15 June 1887, p. 242.

Newton, Alan, 'On Surgical Education: the Halford Oration', *Medical Journal of Australia*, 9 January 1937, p. 42.

Newton, Andrew, The History of the Royal Australasian College of Surgeons from Foundation to 1935, unpublished BA (Hons) thesis, Monash University, 1972.

Nicholas, F. W. and Nicholas, J. M., *Charles Darwin in Australia*, Cambridge University Press, Cambridge, 2002.

O'Sullivan, David M., 'David John Thomas, FRCS, A Founder of Victorian Medicine', *The Medical Journal of Australia*, 30 June, 1956.

Osborne, W. A., 'George Britton Halford: His Life and Work: The Halford Oration' *Medical Journal of Australia*, 19 January 1929, p. 66.

Palmer, Nettie, *Henry Bournes Higgins: A Memoir*, George Harrap, London, 1931.

Payne, Henry (ed), *Proceedings of the First Educational Congress*, State of Victoria, Melbourne, 1912, Mullet, Melbourne.

Pearn, J., 'Hugo Flecker's contributions to medicine and natural history', Department of Child Health, Royal Children's Hospital, Brisbane, Qld. Publication, [no date].

Pensabane, T. S., *The Rise of the Medical Practitioner in Victoria*, Health Research Project, Research Monograph 2, Australian National University, Canberra, 1980.

Porter, Roy, *The Greatest Benefit to Mankind: A Medical History of Humanity from Antiquity to the Present*, Fontana, London, 1999.

Porteus, Stanley D., *A Psychologist of Sorts: The Autobiography and Publications of the Porteus Maze Tests*, Pacific Books, Palo Alto, California, 1969.

Poynter, John and Rasmussen, Carolyn, *A Place Apart: The University: Decades of Challenge*, Melbourne University Press, Carlton South, 1996.

Richardson, Ruth, *Death, Dissection and the Destitute*, Penguin, London, 1989.

Ridley, Ronald (ed), *The Diary of a Vice-Chancellor, University of Melbourne, 1935–1938, Raymond Priestley*, Melbourne University Press, Carlton South, 2002.

Riley, James C., *Rising Life Expectancy: A Global History*, Cambridge University Press, Cambridge, 2001.

Robertson, E. Graeme, 'Melbourne's Public Anatomical and Anthropological Museums, and the Jordans', *Medical Journal of Australia*, 4 February 1956, p. 164.

Roe, Jill, 'Challenge and Response: Religious Life in Melbourne, 1876–86' in *Journal of Religious History*, vol. 5, 1968/9, p. 149.

Roe, Michael, *Nine Australian Progressives: Vitalism in Bourgeois Social Thought 1890–1960*, University of Queensland Press, St. Lucia, 1984.

Russell, Kenneth F., *The Melbourne Medical School, 1862–1962*, Melbourne University Press, Carlton South, 1977.

———. 'Richard James Arthur Berry, 1867–1962', in Geoffrey Kenny and Harold Attwood (eds), *Festschrift for Kenneth Fitzpatrick Russell*, Queensberry Hill Press, Melbourne, 1978.

Schindler, Erwin, 'Arthur Schüller: Pioneer of Neuroradiology', *American Journal of Neuro Radiology*, 18: 1297-1302, August 1997.

Scott, Ernest, *A History of the University of Melbourne*, Melbourne University Press, Melbourne, 1936.

Searle, G. R., *The Quest for National Efficiency: A Study in British Politics and Political Thought, 1899–1914*, Ashfield Press, London, 1990 (first ed. 1971).

Selleck, R. J. W., *Frank Tate: A Biography*, Melbourne University Press, Carlton South, 1982.

———. *The Shop: The University of Melbourne: 1850–1939*, Melbourne University Press, Carlton South, 2003.

Serle, Geoffrey, *Sir John Medley*, Melbourne University Press, Carlton South, 1993.

———. *From Deserts the Prophets Come: The Creative Spirit in Australia, 1788–1972*, Heinemann, Melbourne, 1973.

Smith, F. B., 'Disputes about Typhoid Fever in Victoria in the 1870s', *Health and History, Journal of the Australian Society of the History of Medicine*, vol. 4, no. 2, 2002, p. 1.

Smith, Gordon Keys, Extracts of Memoirs of Dr Gordon Keys Smith, who was a medical student under Professor Wood Jones, 1935–36, and on the staff of the Anatomy Department under Professor Sunderland, 1946–48. The Memoirs 'Reminiscences and Reflections' is an unpublished manuscript, vol. 1, 2005; vol 2 in preparation.

Smyth, R. Brough, *The Aborigines of Victoria*, Ferres, Melbourne, 1878.

Spencer, Frank (ed), *History of Physical Anthropology*, Garland, New York, 1997.

———. *Piltdown: a scientific forgery*, Oxford University Press, London, 1990.

Stawell, Richard, 'The Foundation of a Medical School and the Progress of Medical Education—the Halford Oration' *Medical Journal of* Australia, 3 January 1931, p. 3.

Steele, Edward J., Lindley, Robyn A., Blanden, Robert V., *Lamarck's Signature: How Retrogenes are Changing Darwin's Natural Selection Paradigm*, Perseus Books, 1998.

Stepan, Nancy, *The Idea of Race in Science: Great Britain, 1800–1960*, Macmillan, London, 1982.

Sunderland, Sydney, 'The Melbourne Medical School and some of its "characters" 1931–1975', *Chiron: Journal of the University of Melbourne Medical Society*, vol. 2, no. 2, 1992, p. 46.

The Australasian Students' Song Book, published for the Song Book Board by George Robertson, Melbourne, 1911.

Turnbull, Paul, 'Rare work amongst the professors: the capture of indigenous skulls within phrenological knowledge in early colonial Australia' in Barbara Creed and Jeanette Hoorn (eds), *Body Trade: Captivity, Cannibalism and Colonialism in the Pacific*, Pluto Press, Australia, 2001.

Wilkinson, Ian, 'Frederick McCoy and the University of Melbourne' *The Victorian Naturalist*, vol. 118 (5) 2001, McCoy Special Issue, part 1.

Wood Jones, Frederic, 'The legalization of the study of human anatomy' in J. V. Barry and A. E. Coates (eds), *The Proceedings of the Medico-Legal Society of Victoria during the years 1933, 1934, 1935, 1936*, vol. ii, Brown Prior Anderson Pty Ltd, Melbourne, 1937.

———. *Australia's Vanishing Race*, Angus and Robertson, Sydney, 1934.

———. *Buchanan's Anatomy*, Bailliere, Tindeall and Cox, London, 1943.

———. *Design and Purpose*, Kegan Paul, London 1942.

———. *Habit and Heritage*, Kegan Paul, London 1943.

———. *Hallmarks of Mankind*, Bailliere, Tindeall and Cox, London, 1948.

———. *Life and Living*, Kegan Paul, London, 1939.

———. *Sea Birds Simplified*, Edward Arnold, London, 1934.

———. *Structure and Function as Seen in the Foot*, Bailliere, Tindall and Cox, London, 1944.

———. *The Problem of Man's Ancestry*, SPCK, London, 1918.

———. *Trends of Life*, Edward Arnold, London, 1953.

———. *Unscientific Excursions*, Butler and Tanner, Great Britain, 1934.

Yule, Peter, *The Royal Children's Hospital: a history of faith, science and love*, Halstead Press, Rushcutters Bay, 1999.

Index

Entries are filed word-by-word. **Boldface** page references indicate illustrations; colour plates appearing between pages 126 and 127 are numbered in **Boldface, 126** with page number in parentheses; page references followed by *n* indicate notes.